TIMED ARRAYS

TIMED ARRAYS

Wideband and Time Varying Antenna Arrays

RANDY L. HAUPT

IEEE Press

WILEY

Published by John Wiley & Sons, Inc., Hoboken, New Jersey
Published simultaneously in Canada

For general information on our other products and services or for technical support, please contact our
Customer Care Department within the United States at (800) 762-2974, outside the United States at
(317) 572-3993 or fax (317) 572-4002.

Wiley also publishes its books in a variety of electronic formats. Some content that appears in print may
not be available in electronic formats. For more information about Wiley products, visit our web site at
www.wiley.com.

Library of Congress Cataloging-in-Publication Data:

Haupt, Randy L.
   Timed arrays : wideband and time varying antenna arrays / Randy L. Haupt.
     pages   cm
   Includes index.
   ISBN 978-1-118-86014-4 (cloth)
1. Antenna arrays.   2. Adaptive antennas.   3. Time-domain analysis.   I. Title.
   TK7871.67.A77H383 2015
   621.3841′35–dc23

                             2015004892

Set in 10/12 pt Times by SPi Global, Pondicherry, India

10 9 8 7 6 5 4 3 2 1

To my best friend, mentor, guidance counselor, and wife:
Sue Ellen Haupt and my new granddaughter Adeline.

CONTENTS

LIST OF FIGURES

PREFACE

Antenna arrays have been around for more than a century now. Decades after their introduction, beam steering using phase shifters heralded the introduction of phased arrays. Phased arrays with the help of advancements in computers and electronics morphed into the active electronically scanned array (AESA). As this book shows, a phased array has limited performance, because the design and analysis techniques used are based on phase which is a narrow band concept. High bandwidth signals in a dynamic environment place much higher demands on an antenna array.

This book is an introduction to timed arrays. Phased array is a well-established area with many books already written. Timed array is the next step in which wide bandwidth, large scan sectors, large apertures, and adaptive approaches require an analysis in the time domain rather than the traditional steady-state approach.

The first three chapters lay the foundation for the rest of the book. Arrays are presented from both the usual narrow band approach and from a modulated signal approach. Chapter 4 covers wideband elements and other element issues, such as mutual coupling. Chapter 5 describes feed networks, while Chapter 6 introduces active electronics. An important component of timed arrays is the time delay unit. Chapter 7 describes the need, technology, and placement in the feed network of time delay units. Finally, Chapter 8 introduces different types of adaptive and reconfigurable arrays.

I want to give a special thanks to Dr. Mark Leifer and Dr. Manoja Weiss for their helpful review of portions of this text. Writing a sole-authored book is lonely and challenging. Catching my own errors is difficult. I know there are still some errors lurking about in the text and apologize to the reader in advance.

1

TIMED AND PHASED ARRAY ANTENNAS

This chapter briefly introduces antenna arrays and the difference between phased and timed arrays. Not long after the invention of antenna arrays, researchers experimented with moving the main beam by modifying the phase of the signals fed at the elements [1]. Manual beamsteering eventually led to the invention of the phased array where the main beam was electronically steered to a desired direction by applying a pre-calculated phase to all the elements [2]. Phase is a narrow band concept, though. Today's applications of antenna arrays require high data rates and wide bandwidths. The term "timed arrays" applies to several classes of antenna arrays that are becoming more important with the development of new technologies that must be designed, analyzed, and tested in the time domain, rather than the steady-state, time harmonic forms used with phased arrays. One author defined timed arrays as [3] "… timed-domain equivalent of phased arrays, where each radiating element is excited by pulsed instead of narrowband signals." This book adds adaptive arrays, reconfigurable arrays, and other time-dependent arrays to that definition of timed arrays.

1.1 LARGE ANTENNAS

Large antennas collect more electromagnetic radiation than small antennas. They have the potential to detect very faint signals—the reason they are popular for radio astronomy and radar. Consequently, an antenna has gain that magnifies the received

Timed Arrays: Wideband and Time Varying Antenna Arrays, First Edition. Randy L. Haupt.
© 2015 John Wiley & Sons, Inc. Published 2015 by John Wiley & Sons, Inc.

or transmitted signal in the direction it is pointing. An approximate relationship between the physical size or aperture area (A_p) and gain is given by

$$G = \frac{4\pi A_p}{\lambda^2} \qquad (1.1)$$

where λ is the wavelength. Note that this formula assumes the antenna only transmits over a hemisphere. "Large antenna" may indicate its physical size but more often refers to its electrical size: area$/\lambda^2$. For a given aperture size, higher frequency means higher gain.

Antenna arrays are frequently used in communications and radar systems. The power received by a communications antenna is given by the Friis transmission formula:

$$P_r = \frac{P_t G_t G_r \lambda^2}{(4\pi R)^2} \qquad (1.2)$$

And the power received by a monostatic radar antenna is given by the radar range equation:

$$P_r = \frac{P_t G^2 \lambda^2 \sigma}{(4\pi)^3 R^4} \qquad (1.3)$$

where P_t = power transmitted (W), σ = radar cross section (m^2), R = distance (m^2), λ = wavelength (m), G_t = gain of transmitting antenna, and G_r = gain of receiving antenna.

Both (1.2) and (1.3) point out that a bigger antenna increases the received power through a higher gain.

Large antennas come in two forms: reflectors and arrays. Figure 1.1 has examples of a large reflector (Green Bank Telescope, GBT) used for radio astronomy and a

(a)   (b)

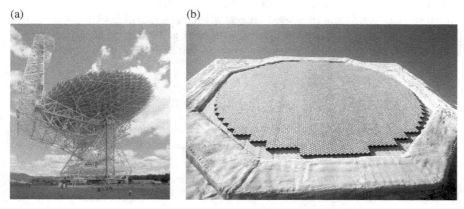

FIGURE 1.1   (a) Green Bank Telescope reflector [4] (Courtesy of NRAO). (b) PAVE PAWS radar [5] (Courtesy of BMDO).

large phased array (PAVE PAWS) used for radar. The Green Bank reflector is a 100 m diameter offset-fed reflector that operates from 0.1 to 116 GHz [4]. The GBT mechanically steers to access 85% of the celestial sphere. The PAVE PAWS radar has two apertures with each covering 240° in azimuth and from 3–85° in elevation [5]. It operates at 420–450 MHz. Each face is 22.1 m in diameter and has 1792 active T/R modules and 885 passive elements resulting in a gain of 37.9 dB with a 2.2° beamwidth. The nominal peak power per face is 600 kW, with a power of 150 kW.

Reflectors and arrays compete to fulfill the role of a large antenna in systems that detect weak signals The reflector is cheap compared to the array, and that is why it is the antenna of choice for commercial activities, such as satellite television. Moving a large reflector in order to locate or track a signal requires gimbals, servomotors, and other mechanical parts that result in a reliability and maintenance problem and leads to significant lifecycle costs [6]. Mechanical steering is too slow for some systems, such as radars on airplanes that need high-speed electronic steering offered by arrays.

An array makes many performance promises for a price. Some potential advantages of an array over a reflector antenna include the following [7]:

1. Fast, wide-angle scanning without moving the antenna
2. Adaptive beamforming
3. Graceful degradation in performance over time
4. Distributed aperture
5. Multiple beams
6. Potential for low radar cross section

Reflectors have the following advantages:

1. High G/T
2. Wide bandwidth
3. Relatively low cost

The mission and budget drive cost/performance trade-offs in the antenna design. Hybrid reflectors are a compromise between the array and reflector. Hybridization may be in the form of arrays of reflectors or an array feed of a reflector. Major research and development efforts are devoted to lowering the cost/ element of an array.

1.2 COLLECTION OF ELEMENTS

Antenna arrays are collections of smaller antenna elements situated to produce a desired gain and response as a function of angle (antenna pattern). Elements either sample (receive) or create sample (transmit) signals. These samples are weighted and then combined to form desired beams. The element configuration determines the spatial sampling rate of the transmit or receive signal.

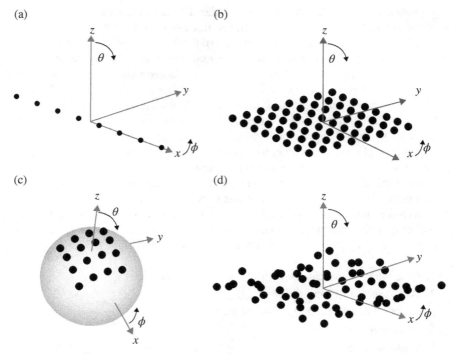

FIGURE 1.2 Four different geometrical layouts of an array: (a) linear array, (b) planar array, (c) conformal array, and (d) random array.

Four common categories of arrays are determined by the element configuration: linear, planar, conformal, or random arrays based on the element layout. The linear array has all of its elements lying along a straight line as shown in Figure 1.2a. Unequally spaced elements lead to an aperiodic or nonuniformly spaced array. The array in Figure 1.2b is planar. The element grid is either periodic or aperiodic. When the elements are placed on a curved surface, then the array is called conformal. A diagram of an array that conforms to a spherical surface is shown in Figure 1.2c. Finally, a random array is shown in Figure 1.2d. The elements in this array follow no pattern in the x, y, and z directions. The minimum element spacing is limited by the area occupied by the element.

Moving far from an antenna creates the illusion that the radiation comes from a single point—much like the light from a star. The phase of the radiated signal on the surface of any sphere having this point as the center is constant, because phase is just distance time the wavenumber, $k = 2\pi/\lambda$. The black dots in Figure 1.2 represent the physical locations of these phase centers of the elements. An element phase center is the point from which the radiation appears to originate (Fig. 1.3). Actual antenna elements occupy a much larger region than the phase centers indicate. These phase centers are also called point sources. In a receive configuration, the point source

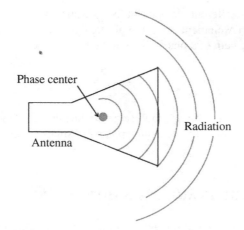

FIGURE 1.3 Concept of an antenna phase center.

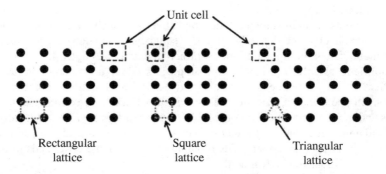

FIGURE 1.4 Three different element lattices and corresponding unit cells.

represents the point in space where the signal sampling occurs. The location of the phase center on an antenna is a function of frequency, polarization, and angle, so placing the phase center at a single point in space is only an approximation.

Elements in a planar array lie in either a rectangular or triangular lattice (Fig. 1.4). A rectangular lattice has equal spacing in the x direction (d_x) that is different from the equal spacing in the y direction (d_y). If the spacing in the x and y directions are the same, then it is a square lattice. Triangular spacing has every other row offset by half the element spacing as shown in Figure 1.4. The triangular lattice is important, because for a given gain (aperture size), it has fewer elements than are needed than for a rectangular lattice [8].

The unit cell is the maximum area that that an element can occupy and is defined by the element spacings

$$A_{uc} = d_x d_y \tag{1.4}$$

Unit cells for the different element lattices are shown in the top row of elements in Figure 1.4. For an N element array, the physical area of the aperture is N times the unit cell area, so the gain of a planar array in (1.1) becomes

$$G = \frac{4\pi N A_{uc}}{\lambda^2} \tag{1.5}$$

Thus, the gain of a planar array is proportional to the number of elements in the array. This formula works when the elements spacing is small enough to prevent grating lobes (see Chapter 3).

1.3 OVERVIEW OF AN ARRAY ARCHITECTURE

Figure 1.5 shows a diagram of the array architecture considered in this book. Other important parts of an array, such as control, cooling, and radome are ignored (although they are very important). The components in Figure 1.5 are critical pieces of either a phased or timed array.

The elements or small antennas that sample the field take various forms. Selection of the antenna element depends on the frequency, polarization, and bandwidth requirements. These components are examined in detail in Chapter 4.

Amplifiers take the form of power amplifiers (PAs) in transmit or low-noise amplifiers (LNAs) in receive. A transmit/receive (T/R) module combines both functions in one package. T/R modules provide the last amplification of transmitted signals and the first amplification of received signals in an array. They typically have phase and amplitude controls to steer the beam, calibrate the signal path, and control the sidelobes. Arrays have many T/R modules, so the module cost needs to be low. Chapter 6 is dedicated to T/R modules.

The beamforming network takes the signals from all the elements and coherently combines them to form a beam in the receive mode. In the transmit mode, the feed

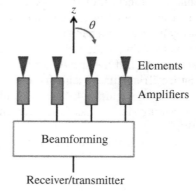

FIGURE 1.5   Generic array architecture.

network distributes the signal from the transmitter to the elements to form a beam. Sometimes, the beamforming network creates multiple beams. For instance, a monopulse tracking radar has simultaneous sum and difference beams. Beamforming is covered in Chapters 3, 5, 7, and 8.

1.4 TRANSIENT VERSUS STEADY STATE

Steady state implies that a system's properties do not change with time. In mathematical terms, the partial derivative of any property with respect to time is zero. Steady-state signals have a finite number of Fourier components contained in a small bandwidth. There are no transients or short duration signals with high bandwidths. A steady-state system is linear time invariant. A steady-state array implies the signals are continuous wave, and the array components do not change with time. Students learn about antenna arrays from this narrow band, steady-state point of view. Even books devoted to phased arrays spend most of their content addressing narrow band, steady-state issues [9–12].

A transient state has quickly changing properties and occurs prior to steady state. Any sudden change in a signal is a *transient*. Transients have an infinite number of Fourier components and a very wide bandwidth. Some books are devoted to one special aspect of phased arrays in the time domain like adaptive nulling or wideband arrays [13, 14].

1.5 TIME VERSUS PHASE

Phase is the narrow band term describing the argument of a sine wave. Phase has units of radians or degrees. Two sinusoids of the same frequency are in phase when their arguments have the same value. Consider a wave described by

$$\cos\left[2\pi f(t + T_{\mathrm{d}}) + \delta_{\mathrm{s}}\right] \tag{1.6}$$

If the frequency (f) remains constant, then the argument of the cosine function depends on the time delay, T_{d}, and the phase shift, δ_{s}. As such, the time delay converts to phase via

$$\delta = 2\pi f T_{\mathrm{d}} \tag{1.7}$$

Since phase is modulo 2π, there is not a direct conversion of phase to time at a given frequency which means that a phase shift converts to an infinite number of possible time delays given by $2\pi f T_{\mathrm{d}}, 4\pi f T_{\mathrm{d}}, 6\pi f T_{\mathrm{d}}, \dots$. This nonreciprocal relationship between time and phase becomes critical when a system is time varying and wideband instead of steady state.

Figure 1.6 shows two sinusoidal signals (steady state) adding in and out of phase. Summing the in-phase signals results in a similar signal having twice the amplitude. Changing the phase of the second signal by $3\pi/4$ moves the two waves relative to each other. Adding these two signals together results in a lower amplitude sinusoid of

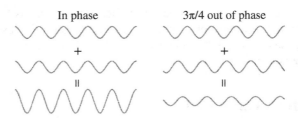

In phase   3π/4 out of phase

FIGURE 1.6   Two signals adding in and out of phase.

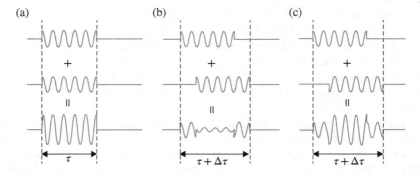

(a)   (b)   (c)

FIGURE 1.7   Pulse dispersion: (a) Two identical pulses (coherent) added together. (b) Two identical pulses (not coherent) that start at different times added together. (c) Two identical pulses (coherent) that start at different times are added together.

the same frequency. The amount of movement of the signal can only be distinguished by $\pm\pi$, even if the phase differential is over 2π. Distinguishing the phase by over 2π is not important for a cosine function, so the difference between time and phase for a sinusoid is not important.

A clear distinction between phase shift and time delay occurs with wideband signals. Adding two identical wide band signals (in this case pulses) together results in a similar signal that has twice the amplitude and has the same start and stop times (Fig. 1.7a). If the start time of the second signal is delayed, then summing the signals causes pulse dispersion (pulse period increases by $\Delta\tau$) as well as destructive interference (Fig. 1.7b). Aligning the phase of both signals before summing them together eliminates the destructive interference but not the dispersion or pulse spreading (Fig. 1.7c). The moral of this story is that phase shifting helps with the coherent addition of two signals but does not help with the pulse dispersion.

1.6   BOOK OVERVIEW

The term "phased array" implies a narrow band approach due to the term "phased." Thus, this book title replaces "phased" with "timed" and concentrates on any aspect of an antenna array that must be viewed from a time perspective. A timed perspective is needed when wideband signals are important. Phase shifters no longer sufficiently align the signals when adding them together. This book explains a variety of issues related to the use of time delay in arrays. Time is also important if the array changes states such as for adaptive arrays or reconfigurable arrays. In these cases, components change array properties in order to meet some performance standard.

The next two chapters introduce signals and the mathematics associated with array theory. From there, a chapter is devoted to each of the array components shown in Figure 1.5. The final chapters describe how to place time delay in an array and introduce adaptive and reconfigurable arrays.

REFERENCES

[1] F. Braun, "Electrical oscillations and wireless telegraphy," Nobel lecture, Dec. 11, 1909.

[2] H. T. Friis and C. B. Feldman, "A multiple unit steerable antenna for short-wave reception," *Proc. IRE*, vol. 25, pp. 841–917, 1937.

[3] G. Franceschetti *et al.*, "Timed arrays in a nutshell," *IEEE Trans. Antennas Propag.*, vol. 53, pp. 4073–4082, 2005.

[4] Wikipedia Contributors, *Green bank telescope*, Wikipedia, The Free Encyclopedia [online]. Available: http://en.wikipedia.org/w/index.php?title=Green_Bank_Telescope&oldid= 629644257. Accessed January 5, 2015.

[5] *Missile Defense Agency* [online]. Available: http://www.mda.mil/global/images/system/ sensors/BealeRADAR3.jpg. Accessed January 5, 2015.

[6] Y. Rahmat-Samii and R. L. Haupt, "Reflector antenna developments: a perspective on the past, present and future," *IEEE Antennas Propagt. Mag.*, vol. 57, 2015.

[7] R. L. Haupt and Y. Rahmat-Samii, "Antenna array developments: a perspective on the past, present and future," *IEEE Antennas Propagt. Mag.*, vol. 57, 2015.

[8] E. Sharp, "A triangular arrangement of planar-array elements that reduces the number needed," *IRE Trans. Antennas Propag.*, vol. 9, pp.126–129, 1961.

[9] R. J. Mailloux, *Phased Array Antenna Handbook* (2nd ed.). Norwood, MA: Artech House, 2005.

[10] R. C. Hansen, *Phased Array Antennas*. Hoboken, NJ: John Wiley & Sons, Inc., 2009.

[11] R. L. Haupt, *Antenna Arrays: A Computational Approach*. Hoboken, NJ: John Wiley & Sons, Inc., 2010.

[12] A. K. Bhattacharyya, *Phased Array Antennas: Floquet Analysis, Synthesis, BFNs and Active Array Systems*. Hoboken, NJ: John Wiley & Sons, Inc., 2006.

[13] A. J. Fenn, *Adaptive Antennas and Phased Arrays for Radar and Communications*. Norwood, MA: Artech House, 2008.

[14] R. A. Monzingo *et al.*, *Introduction to Adaptive Antennas* (2nd ed.). Rayleigh, NC: Scitech Publishing, 2010.

2

RF SIGNALS

Antenna arrays started as very narrow band antennas, in fact continuous wave (CW or single frequency). Historically, this approach worked well, because signals and hardware were narrow band and methods of analyzing and synthesizing arrays were also narrow band and steady state. Many simple analytical expressions exist for narrow band RF systems and work well for pedagogical purposes. In today's world, however, signals are very complex, fast switching, and broadband. Correspondingly, antenna arrays are huge, adaptive, broadband, and complex. The approach to designing and analyzing these new antenna array systems requires numerical simulations as well as experimental measurements that surpass the simpler analytical, steady-state approaches of the past.

This chapter lays the foundation for wide-band signals. Normally, antennas and signals do not appear in the same book, but timed arrays are designed for realistic time-varying signals and scenarios, so the signal is a critical part to understanding the antenna. The first part of this chapter covers the basics of baseband and passband RF signals. The last part of the chapter deals with polarization and bandwidth.

2.1 THE CARRIER AND MODULATION

This section presents two types of signals: a carrier signal that is a high-frequency sinusoid and a lower frequency information (baseband) signal that modulates the carrier. The carrier converts the baseband signal with its low-pass frequency content to

Timed Arrays: Wideband and Time Varying Antenna Arrays, First Edition. Randy L. Haupt.
© 2015 John Wiley & Sons, Inc. Published 2015 by John Wiley & Sons, Inc.

a passband signal with a bandpass spectrum. Selection of the carrier frequency depends on the application.

Electromagnetic theory usually starts with a carrier that is a single-frequency electromagnetic field (Fourier component or tone) represented in rectangular coordinates as follows:

$$\vec{E}(t) = \hat{x}E_x \cos(2\pi f_c t - k_x x) + \hat{y}E_y \cos(2\pi f_c t - k_y y + \psi_y) + \hat{z}E_z \cos(2\pi f_c t - k_z z + \psi_z)$$

$$(2.1)$$

where f_c = carrier frequency; t = time; λ = wavelength; $c = \gamma/T$ speed of light; T = time period; $k = \sqrt{k_x^2 + k_y^2 + k_z^2} = (2\pi/\lambda)$ = wavenumber; $\hat{x}$, $\hat{y}$ and $\hat{z}$ = unit vectors in the x, y, and z directions; E_x, E_y, and E_z = magnitudes of the fields in the x, y, and z directions; and ψ_y and ψ_z = phases of the y and z components relative to the x componant.

The electric field in (2.1) can be written as follows:

$$\vec{E}(t) = \text{Re}\{\mathbf{E}e^{j2\pi f_c t}\}$$

$$(2.2)$$

where $\mathbf{E}$ is the complex steady-state phas or (time independent) given by

$$\mathbf{E} = \hat{x}E_x + \hat{y}E_y e^{j\psi_y} + \hat{z}E_z e^{j\psi_z}$$

$$(2.3)$$

A carrier contains no information and functions only as a means to increase the frequency of the information signal. The baseband (information) signal may be a bit stream or digital signal as shown at the top of Figure 2.1. A high voltage represents a "1" while a low voltage represents a "0" as indicated by "baseband" in Figure 2.1. This baseband signal modulates a single-toned carrier that radiates from the elements.

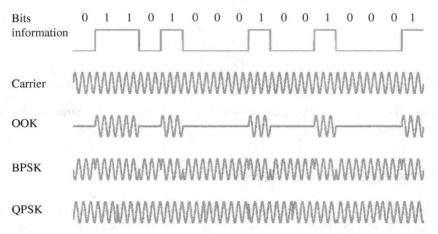

Bits information 0 1 1 0 1 0 0 0 1 0 0 1 0 0 0 1

Carrier

OOK

BPSK

QPSK

FIGURE 2.1 A stream of bits corresponds to a signal amplitude. The baseband signal modulates the carrier. Examples of OOK, BPSK, and QPSK are shown here.

The following three digital modulation schemes convert the bits of the baseband signal to an RF signal:

1. On–off keying (OOK)—A form of amplitude shift keying (ASK) where the amplitude of the baseband signal is proportional to the corresponding bits. The OOK shown in Figure 2.1 multiplies the carrier by the baseband signal and is given by

$$s(t) = \begin{cases} \cos(2\pi f_c t) & b_n = 1 \\ 0 & b_n = 0 \end{cases} \qquad (2.4)$$

2. Binary phase-shift keying (BPSK)—A form of phase modulation in which a change in the baseband from a 1 to a 0 or a 0 to a 1 is indicated by a 180° phase shift. Figure 2.1 shows the constant amplitude BPSK signal. The signal is written as

$$s(t) = \begin{cases} \cos(2\pi f_c t) & b_n = 1 \\ -\cos(2\pi f_c t) & b_n = 0 \end{cases} \qquad (2.5)$$

3. Quadrature phase-shift keying (QPSK)—Bits are grouped into pairs called symbols that correspond to a phase shift of 45, 135, 225, or 315°. The QPSK signal in Figure 2.1 changes phase less frequently than the BPSK signal, because a phase change corresponds to two bits rather than each bit. In the previous two modulation schemes, a symbol is represented by one bit. In QPSK, a symbol requires two bits. A QPSK signal is written as follows:

$$s(t) = \cos[2\pi f_c t + \psi_n], \quad n = 1, 2, 3, 4 \qquad (2.6)$$

where

Bit pair	Phase
00	$\psi_1 = \pi/4$
01	$\psi_2 = 3\pi/4$
10	$\psi_3 = 5\pi/4$
11	$\psi_4 = 7\pi/4$

Many other digital and analog modulation schemes exist, but these three are simple to explain and visualize.

2.2 NOISE AND INTERFERENCE

White noise is an unwanted random signal that has a flat spectrum over the frequency range of interest. If a normal distribution with zero mean describes the noise samples, then the signal is Gaussian white noise. Additive white Gaussian noise (AWGN) means that the Gaussian noise adds to the signal.

Bits   0 1 1 0 1 0 0 0 1 0 0 1 0 0 0 1

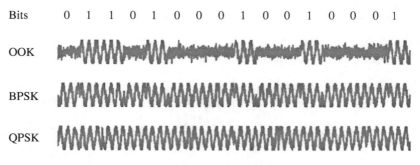

FIGURE 2.2   OOK, BPSK, and QPSK with an SNR of 10 dB.

Characterizing the noise relative to the signal is important and can be done in several ways. The simplest and most common is the signal-to-noise (SNR) ratio:

$$\text{SNR} = \frac{P_s}{P_N} \tag{2.7}$$

where P_s = signal power and P_N = noise power

Figure 2.2 shows the three modulated carriers in Figure 2.1 with an SNR of 10 dB. Phase modulated signals like BPSK and QPSK have greater noise immunity than amplitude modulated signals like OOK, because the phase transitions are less susceptible to the noise than the signal amplitude.

Other signals can intentionally or unintentionally interfere with the desired signal. When these unwanted signals are present, then the signal to interference plus noise ratio (SINR) is specified.

$$\text{SINR} = \frac{P_s}{P_I + P_N} \tag{2.8}$$

where P_I = interference power. The carrier-to-noise ratio (CNR or C/N) is the SNR of a modulated signal. CNR describes the modulated passband signal, while SNR describes the baseband signal after demodulation. Often times, SNR refers to both the passband and base band signals. The bit error rate (BER) is commonly used in digital communications and is the number of bit errors divided by the total number of bits over a long interval.

Intersymbol interference occurs when one symbol or bit overlaps another. This may occur either due to pulse (bit) dispersion (spreading in time) or the signal passing through a bandlimited channel or component. Consider the perfect pulse and its power spectral density (PSD) in Figure 2.3. If the high frequencies are attenuated more than the low frequencies, the pulse becomes rounded and spread in time like the pulse at the bottom left of Figure 2.3. The 3 dB points of the pulse undergoing dispersion have a greater separation than those of the perfect pulse. Dispersion occurs when a pulse, or in our case a bit, increases its width due to either frequency-dependent velocity or attenuation, multipath, or different path lengths between elements in an array. High-frequency attenuation occurs in array components and propagation through some media.

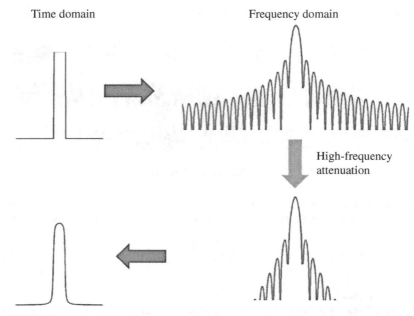

FIGURE 2.3   Pulse dispersion or spreading occurs when high frequencies are attenuated.

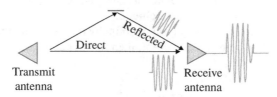

FIGURE 2.4   When the direct and attenuated reflected signal add together, the pulse is dispersed.

Multipath occurs when a signal arrives at a destination via several routes. The paths can be in free space or through transmission lines. If the two paths are exactly of the same length, then the two pulses coherently add together. If one path is longer than the other, then the two pulses add together to create a longer pulse (Fig. 2.4). Figure 2.4 also shows the reflected pulse was attenuated more than the direct pulse.

The BER for the three modulation methods in the presence of white Gaussian noise is approximately given by [1]

$$\text{BER}_{\text{OOK}} = 0.5\text{erfc}\left(\frac{1}{\sqrt{2}}\sqrt{\frac{E_{\text{b}}}{N_0}}\right) \tag{2.9}$$

$$\text{BER}_{\text{BPSK}} = \text{BER}_{\text{QPSK}} = 0.5\text{erfc}\left(\sqrt{\frac{E_{\text{b}}}{N_0}}\right) \tag{2.10}$$

while the symbol error rate for QPSK is given by [1]

$$\text{SER} = 1 - (1 - \text{BER})^2 \qquad (2.11)$$

E_b/N_0 is the energy per bit (E_b) divided by the noise power spectral density (N_0), in other words the SNR per bit. It is a unitless quantity and is frequently used in digital communications to evaluate a communications link.

2.3 POLARIZATION

The polarization of the carrier describes how the magnitude and orientation of the electric field vector changes as a function of time. If the carrier propagates in the z-direction, then the electric field vector lies in the x-y plane. As such, (2.1) becomes

$$\vec{E}(t) = E_{x0} \cos(\omega t - kz)\,\hat{\mathbf{x}} + E_{y0} \cos(\omega t - kz + \Psi_y)\,\hat{\mathbf{y}} \qquad (2.12)$$

If $z = 0$, then (2.12) reduces to

$$\mathbf{E} = E_x \hat{\mathbf{x}} + E_y \hat{\mathbf{y}} \qquad (2.13)$$

where

$$E_x = E_{x0} \cos(\omega t) \qquad (2.14)$$

$$E_y = E_{y0} \cos(\omega t + \Psi_y) \qquad (2.15)$$

The wave in (2.13) is elliptically polarized because it traces an ellipse over time at $z = 0$. The electric field vector rotates in a clockwise or a counter clockwise direction. If your right thumb points in the direction of wave propagation, then your fingers curl in the direction of the E field trajectory ($180° < \Psi_y < 360°$). In this case, the wave is right-hand polarized (RHP). An electric field rotating in the opposite direction (clockwise or $0° < \Psi_y < 180°$) is a left-hand polarized (LHP). The relative phase determines the handedness of the wave.

A polarized wave has an axial ratio (AR) defined by the length of the major axis divided by the length of the minor axis. It ranges from 1 for a circle to ∞ for a line and is often expressed in decibel. The axial ratio is positive for RHP and negative for LHP. *Linear polarization* ($\Psi_y = 0$ and $AR = \infty$) traces a straight line with $E_{x0} = 0$ defining y-polarization (vertical) and $E_{x0} = E_{y0}$ defining x-polarization (horizontal). Slant polarization is a combination of x- and y-polarized waves with zero phase between them.

Circular polarization occurs when the length of the major axis equals the minor axis ($E_{x0} = E_{y0}$, $\Psi_y = +90°$ and $AR = 1$). If $E_{x0} = E_{y0}$, $\Psi_y = +90°$, then the wave is left-hand circularly polarized (LHCP); while if $E_{x0} = E_{y0}$, $\Psi_y = -90°$, then the wave is right-hand circularly polarized (RHCP). Polarization is a very important property of an electromagnetic wave and is usually ignored in communications and digital signal processing books.

Figure 2.5 shows examples of the polarization ellipse as a function of E_y/E_x versus Ψ_y. A straight line indicates linear polarization. The two circles are labeled RHCP and LHCP. The left side of Figure 2.6 is right-hand polarization, while the left side is left-hand polarization. Figure 2.6 is an example of a circularly polarized Gaussian pulse at 1 GHz. The electric field spins in a circle while it first increases then decreases as a function of time.

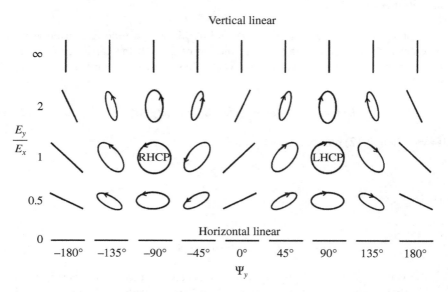

FIGURE 2.5 Wave polarization.

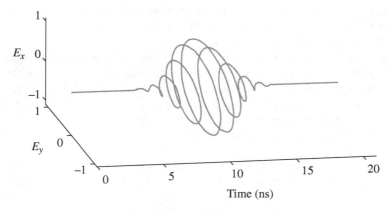

FIGURE 2.6 CP Gaussian pulse.

2.4 SIGNAL BANDWIDTH

The carrier frequency determines which frequency band the system operates (Table 2.1). Since the baseband signal modulates the carrier, the signal occupies a region within one or more of the frequency bands. Bandwidth is defined as the difference between the highest frequency and the lowest frequency in a signal

$$BW = f_{hi} - f_{lo} \qquad (2.16)$$

Often, the system/antenna bandwidth is expresses as a percentage of the carrier frequency:

$$BW = \frac{f_{hi} - f_{lo}}{f_{center}} \times 100 \qquad (2.17)$$

or as a ratio of high to low frequency:

$$BW = \frac{f_{hi}}{f_{lo}} \qquad (2.18)$$

where the highest frequency, f_{hi}, and the lowest frequency, f_{lo}, are determined by some specified rule. The baseband bandwidth of an unmodulated signal is the upper cutoff frequency. Passband bandwidth refers to the signal bandwidth of the modulated signal. Higher frequencies suffer from higher attenuation but have a higher bandwidth potential. Lower frequencies suffer less attenuation but exist in congested frequency bands and have less bandwidth potential.

The power spectral density (PSD) of a signal has units of power/Hertz and shows the amplitude of the frequency components contained in the signal. Analytical

TABLE 2.1   Frequency Bands

Band	Range
HF	3–30 MHz
VHF	30–300 MHz
UHF	300–1000 MHz
L	1–2 GHz
S	2–4 GHz
C	4–8 GHz
X	8–12 GHz
Ku	12–18 GHz
K	18–27 GHz
Ka	27–40 GHz
V	40–75 GHz
W	75–110 GHz
G	110–300 GHz

expressions for the baseband PSD for the three normalized modulation schemes described earlier are given by [1]

BPSK

$$P_{\text{SBPSK}} = T_{\text{s}}\text{sinc}^2\,(\pi f T_{\text{s}})$$ (2.19)

OOK

$$P_{\text{SOOK}} = \frac{1}{2}\left[\delta\,(f) + T_{\text{s}}\text{sinc}^2\,(\pi f T_{\text{s}})\right]$$ (2.20)

QPSK

$$P_{\text{SQPSK}} = 2T_{\text{s}}\text{sinc}^2\,(2\pi f T_{\text{s}})$$ (2.21)

The power spectral density of a passband signal is the PSD shifted to the carrier frequency

$$P_{\text{PASS}} = \frac{1}{4}\left[P_{\text{s}}\,(f - f_{\text{c}}) + P_{\text{s}}\,(f + f_{\text{c}})\right]$$ (2.22)

where P_{s} is given by (2.19), (2.20), or (2.21). The bandwidth of these signals is easy to calculate by setting the P_{s} equal to 0.5, solving for f, then multiplying by 2.

REFERENCE

[1] L. W. Couch II, *Digital and Analog Communication Systems*. New York: MacMillan, 1993.

3

ARRAYS OF POINT SOURCES

Arranging the elements in an aperture is the critical first step in the array design process. Point sources replace real antenna elements in order to determine the element spacing, lattice, aperture shape, and number of elements that meet array performance requirements. Some concepts are easier to explain in terms of a transmit antenna and other times as a receive antenna, so the presentation in this book switches between the two. The type of element, polarization, mutual coupling feed network, and so on are important and will be addressed in subsequent chapters.

3.1 POINT SOURCES

An isotropic point source is a single point in space that radiates equally in all directions. The amplitude and phase of the electromagnetic field radiated by an isotropic point source are constant at a set instant in time and distance from the source. The mathematical representation of the point source in rectangular coordinates is

$$\delta(x)\delta(y)\delta(z) \tag{3.1}$$

This point source is at $(x,y,z) = (0,0,0)$ but can be moved to (x',y',z') by modifying (3.1) to

$$\delta(x-x')\delta(y-y')\delta(z-z') \tag{3.2}$$

Timed Arrays: Wideband and Time Varying Antenna Arrays, First Edition. Randy L. Haupt.
© 2015 John Wiley & Sons, Inc. Published 2015 by John Wiley & Sons, Inc.

The field propagating away from the point source changes according to

$$\frac{e^{j(\omega t - kR)}}{4\pi R} \tag{3.3}$$

while a field propagating toward the point source changes according to

$$\frac{e^{j(\omega t + kR)}}{4\pi R} \tag{3.4}$$

where

$$R = \sqrt{(x - x')^2 + (y - y')^2 + (z - z')^2}$$

$$k = \frac{2\pi}{\lambda}$$

Although the phase of a point source is always isotropic, the amplitude does not have to be isotropic. Elements in an array have gain, and this means the amplitude of the radiated field is a function of angle called the "element pattern." An antenna pattern can be assigned to a point source that is representative of the real element pattern.

3.2   FAR FIELD

The wave radiating from the phase center of an antenna has a constant phase over a sphere of diameter R just like a point source. When this spherical wave impinges on a receive antenna, as shown in Figure 3.1, the wave arrives at the center prior to the edges by $\Delta R/c$ seconds. This spherical wave approximates a plane wave (plane of constant amplitude and phase) when the maximum phase deviation across the aperture is less than $\lambda/16$ or $\pi/8$ radians, which occurs at a distance of [1]

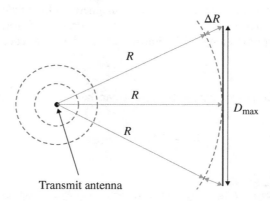

FIGURE 3.1   Antenna far field definition.

$$R = \frac{2D_{max}^2}{\lambda} \tag{3.5}$$

where R is the separation distance between the transmit and receive antennas, and D_{max} is the larger of the maximum dimension of the receive antenna and transmit antenna. Low sidelobe antennas need a larger separation distance, because ΔR must be smaller due to the higher phase accuracies associated with low sidelobe levels. For most applications, an antenna is in the far field, so a plane wave accurately describes the transmit and receive signals.

The Institute of Electrical and Electronics Engineers (IEEE) definition of the far field is for a single frequency. An alternative definition might be derived from a time domain signal. For instance, if ΔR represents 1/16th the distance traveled by the signal in the time represented by one pulse or one bit, then

$$R^2 + \left(\frac{D_{max}}{2}\right)^2 = \left(R + \frac{\tau c}{16}\right)^2$$

$$R \simeq \frac{2D_{max}^2}{\tau c} \tag{3.6}$$

The IEEE selection of $\Delta R = \lambda / 16$ as the standard for the far field is somewhat arbitrary. Selecting $\Delta R = \tau c / 16$ equates (3.5) and (3.6).

3.3   ARRAY SAMPLING IN THE TIME DOMAIN

A periodic linear array along the x-axis has point sources at

$$\delta(x - x_n)\delta(y)\delta(z) \tag{3.7}$$

where

$$x_n = (n-1)d_x \tag{3.8}$$

and d_x is the element spacing. A signal, $s(t)$, incident on the array at an angle θ_s is shown in Figure 3.2. If the signal arrives at element N when $t=0$, then the signal at element n is given by

$$s\left[t - \frac{x_n}{c}\sin\theta_s\right] \tag{3.9}$$

The signal from each element is weighted (w_n), time delayed, and summed to get the array output.

$$S = \sum_{n=1}^{N} w_n s\left[t - \frac{x_n}{c}\sin\theta_s + \tau_n\right] \tag{3.10}$$

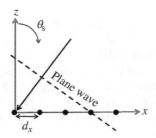

FIGURE 3.2 Plane wave incident on linear array.

If the time delay at element n is

$$\tau_n = \frac{x_n}{c}\sin\theta_s \qquad (3.11)$$

then the array output is a maximum which corresponds to the main beam pointing in the direction of the signal, θ_s. Steering the main beam in this manner is known as time delay steering. The time delay in (3.11) is a function of the distance the signal travels $(x_n\sin\theta_s)$ and the velocity of propagation but is independent of frequency.

For an arbitrary array in three-dimensional space, (3.10) is written as follows:

$$S = \sum_{n=1}^{N} w_n s\left[t - (x_n\sin\theta_s\cos\phi_s + y_n\sin\theta_s\sin\phi_s + z_n\cos\theta_s)/c + \tau_n\right] \qquad (3.12)$$

A planar array has $z_n = 0$. Most planar arrays have elements arranged periodically in the x- and y-directions. Conformal arrays mold to a curved surface, so the element locations have x, y, and z values.

3.4   ARRAY SAMPLING IN THE FREQUENCY DOMAIN

The output from a linear array that receives a CW signal is given by the array factor

$$\mathrm{AF} = \sum_{n=1}^{N} w_n e^{jk(n-1)d_x\sin\theta + \delta_n} \qquad (3.13)$$

Time is ignored in this steady-state scenario. If

$$\delta_n = -(n-1)kd_x\sin\theta_s \qquad (3.14)$$

then the array output is a maximum at θ_s which corresponds to the main beam pointing in the direction of the signal. Steering the main beam in this manner is known as phase steering. A more generalized form of (3.13)

$$\mathrm{AF} = \sum_{n=1}^{N} w_n e^{j\psi_n} \qquad (3.15)$$

where

$$\psi_n = k(x_n \sin\theta \cos\phi + y_n \sin\theta \sin\phi + z_n \cos\theta) + \delta_n \qquad (3.16)$$

and

$$(x_n, y_n, z_n) = \text{location of element } n$$

$$\delta_n = -k(x_n \sin\theta_s \cos\phi_s + y_n \sin\theta_s \sin\phi_s + z_n \cos\theta_s) = -k(x_n u_s + y_n v_s + z_n w_s)$$

$$(\theta_s, \phi_s) = \text{elevation, azimuth steering angles}$$

Uniform arrays have constant element spacing, amplitude weights, and a linear phase distribution for beam steering. A uniform linear array has a closed-form representation of its array factor given by

$$\text{AF}_N = \frac{\sin(N\psi/2)}{N\sin(\psi/2)} \qquad (3.17)$$

A $N_x \times N_y$ rectangular array with elements in a rectangular lattice is defined by

$$x_n = \underbrace{[1 \quad 1 \quad \cdots \quad 1]}_{N_y}^T \underbrace{[1 \quad 1 \quad \cdots \quad 1]}_{N_x} d_x$$

$$y_n = d_y \underbrace{[1 \quad 1 \quad \cdots \quad 1]}_{N_y}^T \underbrace{[1 \quad 1 \quad \cdots \quad 1]}_{N_x} \qquad (3.18)$$

and has the uniform array factor:

$$\text{AF} = \frac{\sin\left(\dfrac{N_x \psi_x}{2}\right)}{N_x \sin\left(\dfrac{\psi_x}{2}\right)} \frac{\sin\left(\dfrac{N_y \psi_y}{2}\right)}{N_y \sin\left(\dfrac{\psi_y}{2}\right)} \qquad (3.19)$$

where $\psi_x = kd_x(u - u_s)$, $\psi_y = kd_y(v - v_s)$, $u_s = \sin\theta \cos\phi$, and $v_s = \sin\theta \sin\phi$

3.5 GRATING LOBES: SPATIAL ALIASING

Aliasing occurs when the array takes less than 2 samples of the signal per period (T in time and λ in space), where

$$\frac{\lambda}{T} = c \qquad (3.20)$$

As a result, the elements must be separated at most by half a wavelength to avoid under sampling an incident plane wave for any angle of incidence. A plane wave

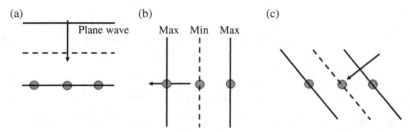

FIGURE 3.3 Array sampling of plane wave from broadside and endfire directions (a) $\theta_s = 0°$, (b) $\theta_s = 90°$, and (c) θ_s arbitrary.

incident at broadside, $\theta_s = 0°$, hits all the elements simultaneously (Fig. 3.3a), so element spacing or spatial sampling is not important. Figure 3.3.b has the wave incident at endfire, $\theta_s = 90°$, which is the worst-case scenario for sampling. As long as the element spacing is $\lambda/2$ at the highest frequency, there is no aliasing. In between broadside and endfire incidence, the Nyquist rate requires that

$$2d \sin \theta_s = \lambda \tag{3.21}$$

Spatial aliasing produces large pattern lobes that steal gain from the main beam and point in directions other than the desired direction. The undesired lobes are grating lobes and for a linear array along the x-axis, their peaks are located at

$$\sin \theta_g = \sin \theta_s \pm \frac{m\lambda}{d_x} \tag{3.22}$$

This equation shows that increasing the frequency moves grating lobes closer to the main beam at θ_s as well as spawning additional grating lobes farther from the main beam. When a grating lobe emerges, the main beam directivity decreases as shown in Figure 3.4.

The array factor for rectangular spacing in a planar array steered to (u_s, v_s) or (θ_s, ϕ_s) is given by

$$AF = \sum_{n=1}^{N} w_n e^{jk\{x_n(u-u_s)+y_n(v-v_s)\}} \tag{3.23}$$

Grating lobes occur at (u_p, v_q) when

$$u_p = u_s + p(1 + \sin \theta_{max\ scan\ x})$$
$$v_q = v_s + q(1 + \sin \theta_{max\ scan\ y}) \tag{3.24}$$

when

$$(u_p - u_s)^2 + (v_q - v_s)^2 \leq 1 \tag{3.25}$$

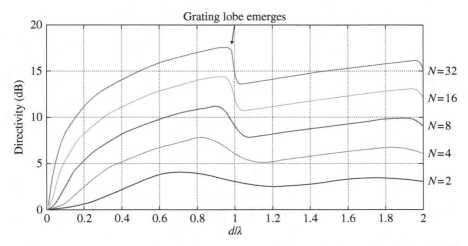

FIGURE 3.4   Directivity of a linear uniform array as a function of element spacing for five different array sizes.

If the element spacing is $d_x = 1.0\lambda$ and $d_y = 1.0\lambda$, then grating lobes appear at $\theta = 90°$ and $\phi = 0°, 90°, 180°, 270°$ when $\theta_s = 0°$. The maximum grating-lobe-free element spacing for an array is based on the maximum angle, $\theta_{\text{max scan}}$ when $m = 1$ and $\theta_g = 90°$ in (3.26)

$$d_x = \frac{\lambda}{1 + \sin\theta_{\text{max scanx}}} \text{ and } d_y = \frac{\lambda}{1 + \sin\theta_{\text{max scany}}} \quad (3.26)$$

This element spacing derived from the grating lobe formula is more restrictive than one that is derived from (3.25).

The d_x for triangular spacing [2] is given in (3.30) with $d_y = d_x\sqrt{3}/2$ and every other row indented by $d_x/2$. Grating lobes for triangular spacing appear at

$$\left(u_s + \frac{p\lambda}{2d_x}, v_s + \frac{q\lambda}{2d_y}\right) \text{ and } \left(u_s + \frac{(2p-1)\lambda}{2d_x}, v_s + \frac{(2q-1)\lambda}{2d_y}\right) \quad (3.27)$$

$$\text{for } p = 0, \pm1, \pm2, \dots \text{ and } q = 0, \pm1, \pm2, \dots$$

When $d_x = 2/\sqrt{3}\lambda = 1.15\lambda$ and $\theta_s = 0°$ six grating lobes appear at $\theta = 90°$ and $\phi = 30, 90, 150, 210, 270$, and $330°$. Equilateral triangular spacing has 13.4% fewer elements than a square grid over the same aperture size for a grating lobe free array factor. Figure 3.5 plots the directivity of a 10×10 array with square and triangular element lattices. The first grating lobe emerges at a higher frequency for the triangular lattice as compared to the square lattice.

Planar arrays usually have a square, rectangular, circular, or hexagonal perimeter. The element spacing and not the array shape determines the appearance of grating lobes. Array shape does play an important role in determining beamwidth, beam shape, and sidelobe levels [3].

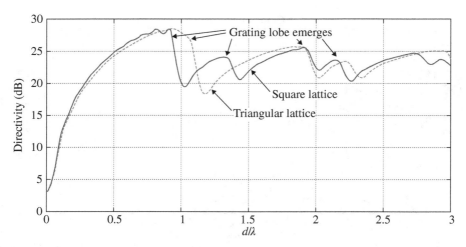

FIGURE 3.5   Directivity of a 10×10 planar uniform array as a function of element spacing for square and triangular element lattices.

3.6   SUBARRAYS AND PANELS

Subarrays are smaller collections of elements in a larger array that have a feed network with an output that combines the signals of the elements in the subarray. In essence, a subarray is an array itself. Elements in a subarray are an electrical grouping—signals are combined to form one signal. A mechanical grouping, called a panel, results from manufacturing size limitations such as PCB board size, packaging constraints, weight, and so on. A large array may consist of smaller panels of elements that "snap" together to form a large array. An electrical grouping is due to the feeding network design or elements that share common components. Usually, they are powers of 4 for a planar array. Figure 3.6 shows a linear array with M levels of subarrays where $N = 2^M$. The signal divider/combiner at each subarray port can have some mechanism for modifying the signal (γ_{mn}), such as an amplitude weight (a_{mn}), phase shifter ($e^{j\delta_{mn}}$), or time delay unit ($e^{j2\pi f \tau_{mn}}$), where m is the subarray level and n is the port number. The total signal weighting at level m and port n is

$$\gamma_{mn} = a_{mn} e^{j\delta_{mn}} e^{j2\pi f \tau_{mn}} \tag{3.28}$$

At level 1 (the element level), there are N ports with N weights.

For computational purposes, all the values of the weightings need to be propagated to the element level. For instance, if γ_{mn} are amplitude weights, then the vector of effective amplitude weights at the elements is given by the $1 \times N$ vector

$$\mathbf{w} = \gamma_1 \circ \gamma_2 \circ \cdots \circ \gamma_M \tag{3.29}$$

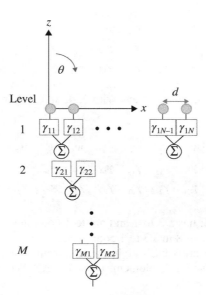

FIGURE 3.6 Power of 2 subarrays in a linear array.

where the MATLAB command

$$\gamma_m = \text{reshape}\left(\text{ones}\left(2^{m-1},1\right)\left[\gamma_{m1}\,\gamma_{m2}\cdots\gamma_{mN/2^{m-1}}\right],1,N\right),\ m=2,3,\ldots,M \quad (3.30)$$

defines a $1 \times N$ vector containing the subarray weight contribution at the element and $\circ$ is element multiplication of two vectors. For an 8-element linear array modeled after Figure 3.6, the γ_m are

$$
\begin{aligned}
\mathbf{w} &= \begin{bmatrix} w_1 & w_2 & w_3 & w_4 & w_5 & w_6 & w_7 & w_8 \end{bmatrix}\\
\boldsymbol{\gamma}_1 &= \begin{bmatrix} \gamma_{11} & \gamma_{12} & \gamma_{13} & \gamma_{14} & \gamma_{15} & \gamma_{16} & \gamma_{17} & \gamma_{18} \\
\boldsymbol{\gamma}_2 = \gamma_{21} & \gamma_{21} & \gamma_{22} & \gamma_{22} & \gamma_{23} & \gamma_{23} & \gamma_{24} & \gamma_{24} \\
\boldsymbol{\gamma}_3 = \gamma_{31} & \gamma_{31} & \gamma_{31} & \gamma_{31} & \gamma_{32} & \gamma_{32} & \gamma_{32} & \gamma_{32} \end{bmatrix}
\end{aligned}
\quad (3.31)
$$

where the effective weight at element 1 is given by

$$w_1 = \gamma_{11}\gamma_{21}\gamma_{31} \quad (3.32)$$

If the number of elements in a subarray is not constant at a given level or the number of elements in a subarray is not a power of 2 at each level, then an alternative formulation to (3.34) is needed [4]. One example is an 8-element array with 2 subarrays of 4 elements each. This arrangement has two 4 to 1 combiners. The effective weights are found from

$$\mathbf{w} = \begin{bmatrix} w_1 & w_2 & w_3 & w_4 & w_5 & w_6 & w_7 & w_8 \end{bmatrix}$$
$$\boldsymbol{\gamma}_1 = \begin{bmatrix} \gamma_{11} & \gamma_{12} & \gamma_{13} & \gamma_{14} & \gamma_{15} & \gamma_{16} & \gamma_{17} & \gamma_{18} \end{bmatrix}$$
$$\boldsymbol{\gamma}_2 = \begin{bmatrix} \gamma_{21} & \gamma_{21} & \gamma_{21} & \gamma_{21} & \gamma_{22} & \gamma_{22} & \gamma_{22} & \gamma_{22} \end{bmatrix} \tag{3.33}$$

If the 8-element array has 3 subarrays with 3, 4, and 1 elements, then the effective weights are given by

$$\mathbf{w} = \begin{bmatrix} w_1 & w_2 & w_3 & w_4 & w_5 & w_6 & w_7 & w_8 \end{bmatrix}$$
$$\boldsymbol{\gamma}_1 = \begin{bmatrix} \gamma_{11} & \gamma_{12} & \gamma_{13} & \gamma_{14} & \gamma_{15} & \gamma_{16} & \gamma_{17} & \gamma_{18} \end{bmatrix}$$
$$\boldsymbol{\gamma}_2 = \begin{bmatrix} \gamma_{21} & \gamma_{21} & \gamma_{21} & \gamma_{22} & \gamma_{22} & \gamma_{22} & \gamma_{22} & \gamma_{23} \end{bmatrix} \tag{3.34}$$

In this example, level 2 has a 3 to 1 and a 4 to 1 combiner whose outputs combine with the output of element 8 in a 3 to 1 combiner.

Assume that a linear array has weights at each element as well as at one level that has N_s subarrays each having N_e elements ($N = N_s \times N_e$). When element n is in subarray p at subarray level m, then

$$w_n = \underbrace{a_{1n} a_{mp}}_{\text{Amplitude}} \underbrace{e^{j\delta_{1n}} e^{j\delta_{mp}}}_{\text{Phase}} \underbrace{e^{j2\pi f \tau_{1n}} e^{j2\pi f \tau_{mp}}}_{\text{Time delay}} \tag{3.35}$$

If the element weights are uniform, $a_{1n} = 1$, then an effective element weight equals the subarray weight at level m and the equation for the linear array factor simplifies to

$$AF = \frac{\sin(N_e k d \sin\theta/2)}{\sin(k d \sin\theta/2)} \sum_{p=1}^{N_s} \gamma_{mp} e^{jk[p-(N_s+1)/2]dN_e \sin\theta} \tag{3.36}$$
$$= AF_e \times AF_s$$

where, AF_e = array factor due to a single subarray and AF_s array factor due to subarray weighting alone.

Equation (3.40) is the product of a uniform subarray factor and the array factor due to the weighted sum of the phase centers of all the subarrays. If $dN_e > \lambda$, then grating lobes due to weight quantization appear at

$$\theta_g = \sin^{-1}\left(\frac{g\lambda}{dN_e}\right) \text{ for } g = 1, 2, \ldots \text{ and } g \le \frac{d}{\lambda} N_e \tag{3.37}$$

The grating lobes can be mitigated by using subarrays of different sizes [4], optimizing the element and subarray weights [5], or overlapping the subarrays [6]. An overlapped subarray has elements that are common to multiple subarrays. A similar, albeit more complicated, subarray analysis can be applied to planar arrays as well [3].

3.7 ELECTRONIC BEAM STEERING

A phase shift only delays a signal by up to one wavelength or 2π radians. In order for a linear phased array to reproduce the signal with 100% accuracy, then the phase difference between the signal received at the first and last elements must be less than 2π radians.

$$k\left(x_N - x_1\right)\sin\theta_s \le 2\pi \qquad (3.38)$$

If not, then time delay units are needed. In practice, however, (3.42) is unnecessarily stringent.

One limitation to the bandwidth of a phased array is the actual versus the desired main beam pointing direction at the maximum scan angle as a function of frequency. At frequency, f, the main beam squints (moves) to angle θ from the desired steering angle according to [7]

$$\sin\theta = \frac{f_c}{f}\sin\theta_s \qquad (3.39)$$

where f_c is the center frequency. Figure 3.7 is a plot of the main beam pointing direction as a function of scan angle and frequency. At a given scan angle, the main beam moves toward boresight as the frequency increases and away from boresight as the frequency decreases. Phased arrays have significant beam squint at large scan angles. In some narrow band systems, beam steering is done by changing the frequency and taking advantage of the beam squint.

Figure 3.8 demonstrates beam squint at $\theta_s = 60°$ for a 32-element linear phased array with a center frequency of 2 GHz. At the high frequency (2.1 GHz), the beam points to 55.6° and at the low frequency (1.9 GHz), the beam points to 65.7°. The

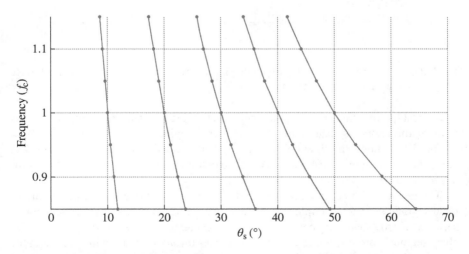

FIGURE 3.7 Beam squint as a function of frequency and steering angle.

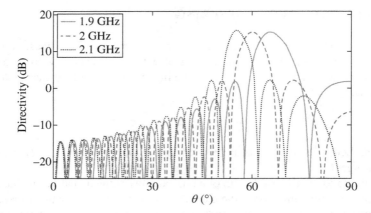

FIGURE 3.8   Beam squint associated with a 32-element phased array when the main beam is steered to 60° at the center frequency.

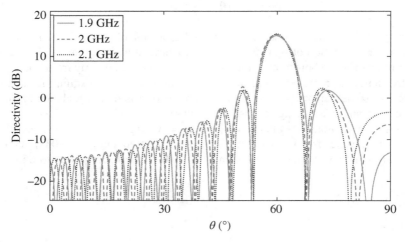

FIGURE 3.9   Array factors at the center, high, and low frequencies of a 32-element timed array when the main beam is steered to 60°.

squint results from the constant phase shift over all the frequencies. Using time delay units at the elements instead of phase shifters eliminates beam squint (Fig. 3.9). Figure 3.10a shows the time delay steering that is translated into phase at 1.8, 2.0, and 2.2 GHz in Figure 3.10b. Unlike the time delay, the slope of the linear phase shift at each frequency is different.

The beam squint can be managed if the wideband signal is divided into narrow band segments. As an example, a wideband linear FM chirp can be generated from a series of narrow-band segments of the signal centered at different frequencies. The chirp segments are transmitted as separate pulses through a beam with an acceptable amount of squint caused by the phase shifters [8]. Chirp segments are transmitted

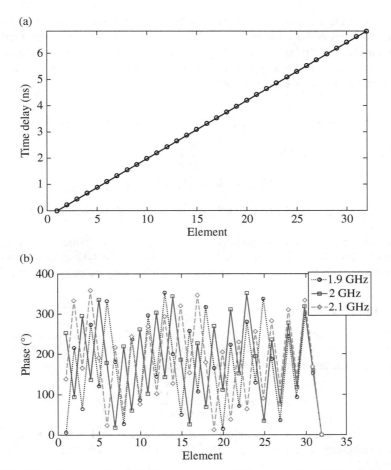

FIGURE 3.10 Beam steering a 32-element array to 60°: (a) time delay (b) time delay translated into phase at three frequencies.

one at a time from the lowest to highest frequency. The image is reconstructed by combining all pulses from all chirp segments. The resulting image has a resolution consistent with the entire resolution bandwidth, which is much larger than any segment's chirp bandwidth.

3.8 AMPLITUDE WEIGHTING

Lowering the sidelobe levels of an array helps the array reject strong interference that enters the sidelobes and degrades the SINR. The binomial amplitude taper theoretically eliminates sidelobes but is not very useful in practice [9]. Schelkunoff developed the idea of controlling the array pattern sidelobes by using the unit circle to move the nulls that in turn change the sidelobes [10]. Dolph built on Schelkunoff's ideas to map the zeros of a Chebyshev polynomial to the nulls of an array factor to get all sidelobes at the same level [11]. More practical implementations were made by Taylor [12, 13]. Today, numerical optimization allows array designers to take into account many practical issues that simple analytical approaches cannot address [14].

A linear array factor transforms into a polynomial

$$AF = \sum_{n=1}^{N} w_n z^{(n-1)} = w_N \left(z^{N-1} + \frac{w_{N-1}}{w_N} z^{N-2} + \frac{w_{N-2}}{w_N} z^{N-3} + \cdots + \frac{w_1}{w_N} \right) \quad (3.40)$$

that factors into product form

$$AF = (z - z_1)(z - z_2) \cdots (z - z_{N-1}) \quad (3.41)$$

where, $z = e^{j\psi}$, $\psi = kd \sin\theta$, $z_n = e^{j\psi_n}$, $\psi_n = kd \sin\theta_n$, $\theta_n =$ location of the nth null, and $w_N = 1$

Moving the nulls or zeros of the polynomial changes the array weights. If the nulls are symmetric about the main beam, then the weights are real. Analytical derivations of low sidelobe amplitude tapers are based on appropriately locating the nulls in (3.45) [15].

3.8.1 Dolph–Chebyshev Taper

As previously mentioned, the Dolph–Chebyshev amplitude taper for a linear array produces an array factor with all the sidelobe peaks at the same level. This taper results when the zeros of (3.45) are calculated using [3]

$$\psi_n = 2\cos^{-1} \left\{ \frac{\cos((n-0.5)\pi/N - 1)}{\cosh(\pi A/N - 1)} \right\} \quad (3.42)$$

where $A = (1/\pi)\cosh^{-1}(10^{sll/20})$ and sll is the sidelobe level in decibel. The 12-element 30 dB Chebyshev amplitude taper in Figure 3.11 produces the array factor in Figure 3.12.

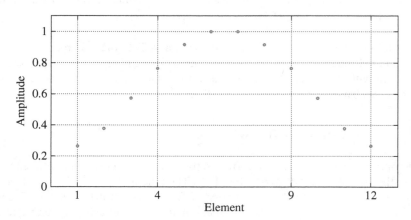

FIGURE 3.11 Amplitude weights for a 30 dB Chebyshev taper.

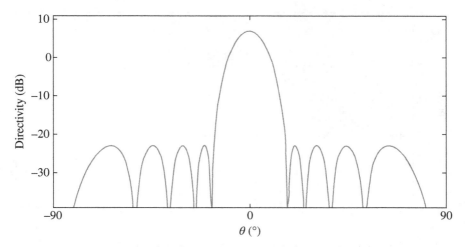

FIGURE 3.12   Array factor for a 30 dB Chebyshev taper.

3.8.2   Taylor Tapers

The Chebyshev weighting is practical for a small linear array, but as N becomes large, other amplitude tapers are better suited for low sidelobe arrays. One such taper developed by Taylor [12] is similar to the Chebyshev taper in that the maximum sidelobe level can be specified. The difference is that the Taylor taper only has the first $\bar{n} - 1$ sidelobes on either side of the main beam at a specified height, while all remaining sidelobes decrease at the same rate as the sidelobe envelope of a uniform array.

As a result, the Taylor taper only moves the first $\bar{n} - 1$ nulls on either side of the main beam away from the main beam. Since the other zeros remain untouched, the outer sidelobes and nulls are identical to those of a uniform array. The null locations for the Taylor array factor are calculated from [3]

$$\psi_n = \frac{\pm 2\pi}{N} \begin{cases} \bar{n} \sqrt{\dfrac{A^2 + (n - 0.5)^2}{A^2 + (\bar{n} - 0.5)^2}} & n < \bar{n} \\[4mm] n & n \geq \bar{n} \end{cases} \tag{3.43}$$

As with the Chebyshev taper, these ψ_n are substituted into (3.45) to find the amplitude weights.

For a specified sidelobe level, there is an $\bar{n}$ that results in a maximum directivity. As $\bar{n}$ increases, more energy goes from the main beam into the sidelobes until a point is reached where the directivity decreases. As $\bar{n}$ decreases, the beamwidth increases, so power is robbed from the peak of the main beam in order to increase the width of the main beam. Although a large taper efficiency, η_T, is desirable, it is not the only consideration when implementing an amplitude taper. For small $\bar{n}$, the amplitude taper monotonically decreases from the center to the edge. The increased amplitude at the edges is undesirable due to mutual coupling. Above a certain $\bar{n}$ for a given sidelobe

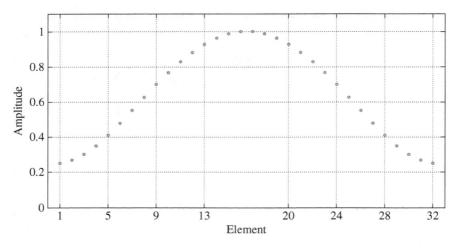

FIGURE 3.13   Amplitude weights for a 30 dB $\bar{n} = 5$ Taylor taper.

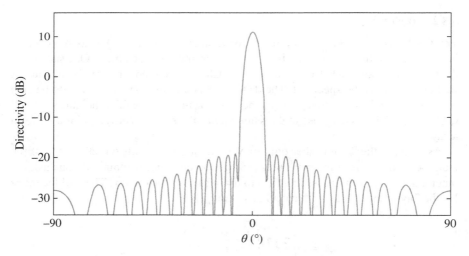

FIGURE 3.14   Array factor for a 30 dB $\bar{n} = 5$ Taylor taper.

level, however, the amplitude taper increases at the edges. The 32-element 30 dB $\bar{n} = 5$ sidelobe Taylor amplitude taper in Figure 3.13 produces the array factor of in Figure 3.14. There is a version of the Taylor taper for circular apertures as well [13].

3.8.3   Bayliss

The Bayliss amplitude taper produces low sidelobes for difference patterns [16]. It has $\bar{n} - 1$ sidelobes at a specified height at symmetric locations from boresight, while the rest decrease away from the main beam as with a uniform difference pattern. Since the left half of the array is 180° out of phase with the right half, a deep null forms at broadside. This null is ideal for locating targets. Null locations are given by [3]

$$\psi_n = \frac{2\pi}{Nd} \begin{cases} 0 & n = 0 \\ (\bar{n}+0.5)\sqrt{\dfrac{q_n}{B^2+\bar{n}^2}} & 1 \le \bar{n} \le 4 \\ (n+0.5)\sqrt{\dfrac{B^2+n^2}{B^2+\bar{n}^2}} & 5 \le \bar{n} \le \bar{n}-1 \\ n & n \ge \bar{n} \end{cases} \tag{3.44}$$

where

$$B = 0.3038753 + \text{sll}\left\{0.05042922 + \text{sll}[-2.7989\times10^{-4} + \text{sll}(3.43\times10^{-6} - 2\times10^{-8}\,\text{sll})]\right\}$$

$$q_1 = 0.9858302 + \text{sll}\left\{0.0333885 + \text{sll}[1.4064\times10^{-4} + \text{sll}(-1.9\times10^{-6} + 1\times10^{-8}\,\text{sll})]\right\}$$

$$q_2 = 2.00337487 + \text{sll}\left\{0.01141548 + \text{sll}[4.159\times10^{-4} + \text{sll}(-3.73\times10^{-6} + 1\times10^{-8}\,\text{sll})]\right\}$$

$$q_3 = 3.00636321 + \text{sll}\left\{0.00683394 + \text{sll}[2.9281\times10^{-4} + \text{sll}(-1.61\times10^{-6})]\right\}$$

$$q_4 = 4.00518423 + \text{sll}\left\{0.00501795 + \text{sll}[2.1735\times10^{-4} + \text{sll}(-8.8\times10^{-7})]\right\}$$

$$\tag{3.45}$$

The 32 element 30 dB $\bar{n} = 5$ sidelobe Bayliss amplitude taper in Figure 3.15 produces the array factor of in Figure 3.16. There is a version of the Bayliss taper for circular apertures as well.

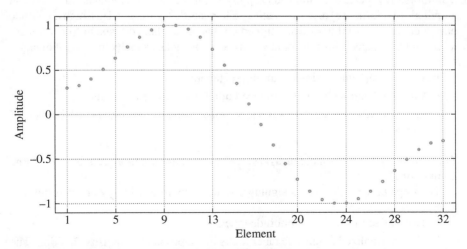

FIGURE 3.15 Amplitude weights for a 30 dB $\bar{n} = 5$ Bayliss taper.

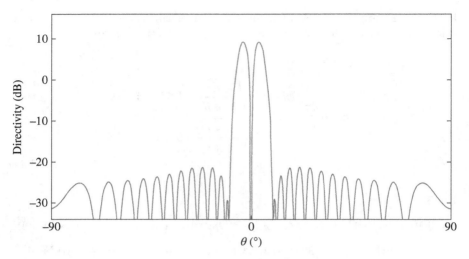

FIGURE 3.16   Array factor for a 30 dB $\bar{n} = 5$ Bayliss taper.

3.9   THINNED ARRAYS

An amplitude taper may also be placed on an array aperture by making some of the elements in a uniform array inactive in a way that the aperture current density mimics a desired amplitude taper across the array aperture [17]. The inactive elements are connected to a matched load and deliver no signal to the beamformer. Inactive elements are not removed from the aperture, so the periodic element lattice is not disturbed and non-edge elements see similar mutual coupling environments.

The normalized desired amplitude taper serves as a probability density function for a uniform array that is to be thinned [18]. In a thinned array, the elements either have an amplitude of one (active) or zero (inactive). Elements that correspond to a high amplitude have a greater probability of being turned on than those that correspond to a low sidelobe amplitude taper. This type of taper has some advantages including the following:

- A cheap implementation an amplitude taper.
- A narrow beamwidth with a reduced number of active elements.
- Each non-edge element sees the same environment, so mutual coupling is well defined.

Thinning works best for large arrays, since the statistics are more reliable for a large number of elements.

The steps in the recipe for designing a statistically thinned array are as follows:

1. Normalize the desired amplitude taper, a_n^{desired}.
2. This normalized amplitude taper looks like a probability density function. The probability that an element is on is equal to its desired normalized amplitude.

3. Generate a uniform random number, r, $0 \le r \le 1$.

4. $w_n = \begin{cases} 1 & \text{if } r \le w_n^{\text{desired}} \\ 0 & \text{if } r > w_n^{\text{desired}} \end{cases}$

5. This process repeats for each element in the array.

Thinning does not result in a symmetric aperture taper.

The total number of elements in the array is the sum of the active (N_a) and inactive (N_i) elements:

$$N = N_a + N_i \qquad (3.46)$$

Array directivity and sidelobe level depend on the number of active elements. As an example, the directivity of a thinned linear array with half wavelength spacing is

$$D_{\text{thin}} = N_a \qquad (3.47)$$

The taper efficiency, η_t, is found by dividing the directivity of the thinned array by the directivity of the array if all the elements were active.

$$\eta_t = \frac{N_a}{N} \qquad (3.48)$$

An expression for the rms sidelobe level of a thinned array is given by [19]

$$\overline{\text{sll}^2} = \frac{\eta_{t,\text{desired}} N - N_a}{\eta_{t,\text{desired}} N N_a} \qquad (3.49)$$

where $\overline{\text{sll}^2}$ is the power level of the average sidelobe level and $\eta_{t,\text{desired}}$ is the taper efficiency of the target amplitude taper. An expression for the peak sidelobe level (sll^2) is found by assuming all the sidelobes are within three standard deviations of the rms sidelobe level and has the form [19]

$$P\left(\text{all sidelobes} < \text{sll}_p^2\right) \approx \left(1 - e^{-\text{sll}_p^2/\overline{\text{sll}^2}}\right)^{N/2} \qquad (3.50)$$

for linear and planar arrays having half wavelength spacing.

A uniform 64×64 element array of point sources having a $\cos\theta$ element pattern in a square grid with half wavelength element spacing has a directivity of 47.1 dB and a peak sidelobe level 13.25 dB below the peak of the main beam. Thinning the aperture to get a 25 dB $\bar{n} = 5$ Taylor taper (Fig. 3.17) in both the x and y directions results in the thinned aperture shown in Figure 3.18. It has 1990 active and 2106 inactive elements. The corresponding array pattern cuts at $\varphi = 0°$ are shown in Figure 3.19. In addition, the rms sidelobe level calculated using (3.49) is on the plot. The main beams and first sidelobes of the two patterns are nearly identical. The thinned array shows more deviations from the desired at higher angle from boresight where the

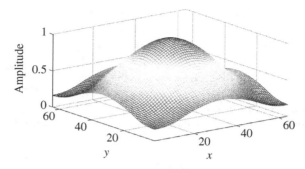

FIGURE 3.17   25 dB Taylor taper for the 64×64 planar array.

FIGURE 3.18   64×64 element array thinned to get a 25 dB $\bar{n} = 5$ Taylor taper.

sidelobes of the desired pattern are lower. Thinning can be optimized to improve the aperture efficiency—more active elements (higher directivity) for the same sidelobe level [20].

Thinning reduces sidelobes and delays the onset of grating lobes while maintaining a periodic grid. Nonuniform spacing or aperiodic arrays, on the other hand, achieve the same objective. Feeding and phasing periodic arrays is usually easier than aperiodic arrays. For large elements in a big array, as with radio telescopes, are often placed in a nonuniform grid. The Long Wavelength Array Station 1 (LWA1) is a radio telescope operating from 10 to 88 MHz [21]. The 256 elements in the array

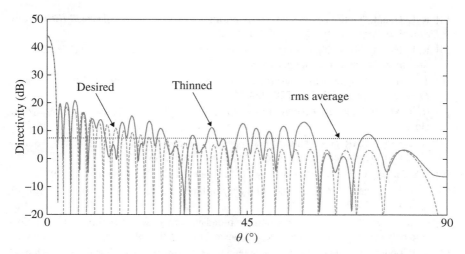

FIGURE 3.19 Solid line is an array factor cut of the 64×64 element array thinned to get a 25 dB $\bar{n} = 5$ Taylor pattern. The dashed line is a cut from the desired low sidelobe pattern. The dotted line is the rms average sidelobe level predicted by (3.49).

FIGURE 3.20 Photograph of the random arrangement of elements in the LWA1 array. Reprinted by permission of Ref. [21]; © 2013 IEEE.

have a mean spacing of 5.4 m with a minimum spacing of 5 m (smallest distance that does not interfere with maintenance) inside an ellipse that has the major axis in the north–south direction (Fig. 3.20).

REFERENCES

[1] IEEE-SA Standards Board, *IEEE Standard for Definitions of Terms for Antennas*, IEEE Standard 145–2013.

[2] E. Sharp, "A triangular arrangement of planar-array elements that reduces the number needed," *IEEE Trans. Antennas Propag.*, vol. 9, pp.126–129, 1961.

[3] R. L. Haupt, *Antenna Arrays: A Computational Approach*. Hoboken, NJ: John Wiley & Sons, Inc., 2010.

[4] R. L. Haupt, "Optimized weighting of uniform subarrays of unequal sizes," *IEEE Trans. Antennas Propag.*, vol. 55, pp.1207–1210, 2007.

[5] R. L. Haupt, "Reducing grating lobes due to subarray amplitude tapering," *IEEE Trans. Antennas Propag.*, vol. 33, pp. 846–850, 1985.

[6] T. Azar, "Overlapped subarrays: review and update [education column]," *IEEE Antennas Propagat. Mag.*, vol. 55, pp. 228–234, 2013.

[7] R. J. Mailloux, *Phased Array Antenna Handbook* (2nd ed.). Norwood, MA: Artech House, 2005.

[8] A. Doerry, "SAR processing with stepped chirps and phased array antennas," Sandia Natl. Lab., Tech. Rep. SAND2006-5855, Albuquerque, NM, 2006.

[9] J. S. Stone, "Directive antenna array," U.S. Patent 1 643 323, September 27, 1927.

[10] S. A. Schelkunoff, "A mathematical theory of linear arrays," *Bell Syst. Tech. J.*, vol. 22, pp. 80–107, 1943.

[11] C. L. Dolph, "A current distribution for broadside arrays which optimizes the relationship between beam width and side-lobe level," *Proc. IRE*, vol. 34, pp. 335–348, 1946.

[12] T. T. Taylor, "Design of line source antennas for narrow beamwidth and low side lobes," *IRE Trans. Antennas Propag.*, vol. 3, pp. 16–28, 1955.

[13] T. Taylor, "Design of circular apertures for narrow beamwidth and low sidelobes," *IEEE Trans. Antennas Propag.*, vol. 8, pp. 17–22, 1960.

[14] R. L. Haupt and D. H. Werner, *Genetic Algorithms in Electromagnetics*. Hoboken, NJ: John Wiley & Sons, Inc., 2007.

[15] R. S. Elliott, *Antenna Theory and Design* (rev. ed.), Hoboken, NJ: John Wiley & Sons, Inc., 2003.

[16] E. T. Bayliss, "Design of monopulse antenna difference patterns with low sidelobes," *Bell Syst. Tech. J.*, vol. 47, pp.623–650, 1968.

[17] R. Willey, "Space tapering of linear and planar arrays," *IEEE Trans. Antennas Propag.*, vol. 10, pp. 369–377, 1962.

[18] M. Skolnik *et al.*, "Statistically designed density-tapered arrays," *IEEE Trans. Antennas Propag.*, vol. 12, pp. 408–417, 1964.

[19] E. Brookner, "Antenna array fundamentals-part 1," in *Practical Phased Array Antenna Systems*, E. Brookner, Eds. Norwood, MA: Artech House, 1991, pp. 2–38.

[20] R. L. Haupt, "Thinned arrays using genetic algorithms," *IEEE Trans. Antennas Propag.*, vol. 42, pp. 993–999, 1994.

[21] S. W. Ellingson, "The LWA1 radio telescope," *IEEE Trans. Antennas Propag.*, vol. 61, pp. 2540–2549, 2013.

4

ELEMENTS IN TIMED ARRAYS

Chapters 1–3 only dealt with arrays of point sources. Point sources are not real antennas and represent a single point from which an antenna appears to radiate. The resulting array factor depends only on the point source spacing, the element weights, and the frequency. A good approximation of an array pattern is the product of the element pattern of a real antenna times the array factor. Elements determine the array bandwidth and polarization and impact the array gain. Spacing between elements limits the size of the elements that fit into the array lattice. The unit cell is the maximum area allotted to an element in the array aperture and is defined by the area $d_x d_y$. The characteristics of an isolated element (impedance, gain, polarization, etc.) change when the element is surrounded by other elements in the array. This chapter introduces potential elements for a timed array and their characteristics.

4.1 ELEMENT CHARACTERISTICS

An array of point sources is a simple model for estimating the array beamwidth, directivity, sidelobe level, and grating lobe formation. Other array characteristics depend on the type of element in the array. The array pattern approximately equals the product of the array factor and the element pattern that combines the characteristics of both the element positions and weightings with the characteristics of the element.

Timed Arrays: Wideband and Time Varying Antenna Arrays, First Edition. Randy L. Haupt.
© 2015 John Wiley & Sons, Inc. Published 2015 by John Wiley & Sons, Inc.

4.1.1 Polarization

An antenna's polarization is defined by the polarization of a wave transmitted by the antenna [1]. The orientation of the time-varying electric field incident on a receive antenna determines the current flow induced in the antenna. Arrays are designed to be either linearly or circularly polarized. A circularly polarized array is the result of circularly polarized elements, linearly polarized elements that are properly phased, or groups of linearly polarized elements that are appropriately rotated and phased. Circular polarization has an axial ratio (AR) of 1, while linear polarization has an AR of ∞. In between those extremes is elliptical polarization. As a rule of thumb, a wave is no longer circularly polarized if the AR is greater than 2 (3 dB). The receive antenna should have the same polarization as the transmit antenna.

A time-varying electric field induces a time-varying current in a wire parallel to the field, no current in a wire perpendicular to the field, and some time-varying current in a wire oriented between parallel and perpendicular. A thin wire dipole has current flowing in one direction, so it is linearly polarized in that direction. The left side of the equal sign in Figure 4.1 shows a horizontal dipole that radiates a horizontally polarized electric field added to a vertical dipole that radiates a vertically polarized electric field. Feeding them 90° out of phase causes circular polarization. A −90° phase produces right-hand circular polarization (RHCP) while a +90° phase produces left-hand circular polarization (LHCP).

Figure 4.2 shows two approaches to getting circular polarization from linearly polarized antennas arranged and phased in a 2 × 2 subarray [2]. In Figure 4.2a, only 90° phase shifts are needed, because the microstrip patches are flipped in order to produce an additional 180° shift. The resulting CP is very poor at angles greater than 5° off broadside, though. The 2 × 2 subarray in Figure 4.2b maintains better CP at larger scan angles than the Figure 4.2a. Figure 4.3a is an example of a CP ultra wideband array made from four linearly polarized horn antennas properly rotated and phased [3]. Arrows drawn on the horns indicate the direction of the electric field. The measured and calculated pulses transmitted at boresight are shown in Figure 4.3b. Its measured AR was less than 3 dB for scan angles up to 21°.

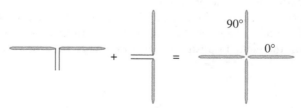

FIGURE 4.1 Combining two linear polarized elements to get circular polarization (in this case, LHCP).

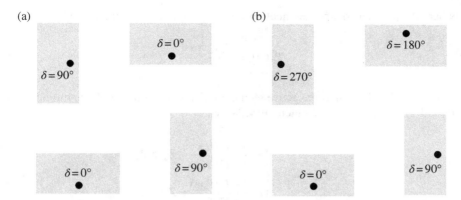

FIGURE 4.2 Four linearly polarized patches are rotated and phased to get circular polarization: (a) phasing and placement for limited scan and (b) phasing and placement for wide scan.

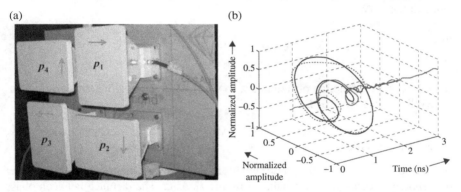

FIGURE 4.3 (a) Four linearly polarized broadband horns are rotated and phased to get circular polarization. (b) Measured (dotted) and calculated (solid) pulses transmitted by array at boresight. Reprinted by permission of Ref. [3]; © 2010 IEEE.

4.1.2 Impedance

The element impedance must match the feed line impedance for maximum power transfer. The reflection coefficient is given by

$$\Gamma = s_{11} = \frac{V_{\text{reflected}}}{V_{\text{incident}}} = \frac{Z_A - Z_C}{Z_A + Z_C} \tag{4.1}$$

where Z_A is the antenna impedance and Z_C is the transmission line impedance. When a large planar array has no grating lobes, then the reflection coefficient at an element when all other elements are terminated in matched loads is approximately [4]

$$\left|\Gamma(\theta,\phi)\right| = \sqrt{1 - \frac{\lambda^2 P(\theta,\phi)}{d_x d_y \cos\theta}} \tag{4.2}$$

where the terminated array element power pattern has a maximum given by

$$P(0,0) = \frac{d_x d_y \cos\theta}{\lambda^2} \tag{4.3}$$

Equation 4.2 is known as the active element reflection coefficient. Substituting (4.2) into (4.1) gives the active element impedance.

$$Z_A(\theta,\phi) = \begin{cases} Z_c \dfrac{1+\Gamma(\theta,\phi)}{1-\Gamma(\theta,\phi)} & Z_A \geq Z_c \\[2mm] Z_c \dfrac{1-\Gamma(\theta,\phi)}{1+\Gamma(\theta,\phi)} & Z_A < Z_c \end{cases} \tag{4.4}$$

The voltage standing wave ratio (VSWR) is the ratio of the maximum to minimum value of the VSW

$$\text{VSWR} = \frac{V_{max}}{V_{min}} = \frac{1+|\Gamma|}{1-|\Gamma|} \tag{4.5}$$

Impedances are a function of frequency, so the VSWR bandwidth is often defined as a VSWR < 2 or $\Gamma < -10\,\text{dB}$.

4.1.3 Phase center

A point source represents the phase center of the antenna element, which is the center of a sphere of constant phase radiated by the antenna. The distance between phase centers of adjacent elements is the element spacing. This phase center moves with frequency and angle; so in actuality, it only exists for a portion of a sphere at a given frequency. Wideband antennas do not have a single-phase center.

4.1.4 Conformal

A conformal array has the element phase centers (x_n, y_n, z_n) lying on a curved surface. Some elements easily conform to a surface, while others do not. Elements that have a significant depth are more difficult to conform to a surface than flat elements. Slight curvatures have little effect on element performance.

4.1.5 Size

An element must fit within the mechanical structure of an array. The element has an area defined by the unit cell and a depth. T/R modules, cooling, control lines, feed lines, and so on compete for space behind the element and determine the depth or thickness of the array. Elements may also have weight limitations.

4.1.6   Directivity

The directivity of an element is related to its size and is approximated by

$$D_{\text{element}} = \frac{4\pi \, dxdy}{\lambda^2} \tag{4.6}$$

Planar array directivity is the sum of the directivities of all N elements [5].

$$D_{\text{array}} = ND_{\text{element}} = \frac{4\pi \, Ndxdy}{\lambda^2} \tag{4.7}$$

4.1.7   Bandwidth

The element bandwidth is typically defined by the input impedance, but other characteristics, such as polarization or gain, may also be used. Impedance bandwidth is the frequency range over which the antenna return loss is less than −10 dB or the VSWR is less than 2. This definition allows the antenna impedance to vary by 22.2% over the bandwidth, while the feed line impedance remains constant. The element bandwidth should operate over the frequencies in the array bandwidth (chapter 3) and should include the mutual coupling effects (Section 4.3), because the element bandwidth in the array is different from the isolated element bandwidth.

CP bandwidth is usually defined by an AR less than 3 dB, which equates to the major axis being twice as long as the minor axis.

The system bandwidth defines all the frequencies over which a signal can be transmitted or received by the array. The system bandwidth may be quite a bit larger than the signal or instantaneous bandwidth. For instance, an array may transmit a narrow band signal (e.g., 10 MHz) at a carrier frequency (e.g., 100 MHz) that can be varied over a wide bandwidth (e.g., 2 GHz).

4.1.8   Balun

A balun controls the current flow between the transmission line (unbalanced) and antenna (balanced). Figure 4.4 is a picture of a coaxial cable feeding a dipole with the current flow indicated. Ideally, I_1 is equal and in the same direction as I_2, and I_4 is equal and in the opposite direction of I_5 with $I_3 = 0$ (the current on the outside of the

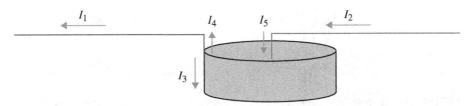

FIGURE 4.4   Currents on a coax-fed dipole antenna.

cable). Unfortunately, the current on the inside of the outer conductor of the coax flows on the outside of the cable as well as to the left arm of the dipole. As a result, the coax radiates, because there is no current that cancels I_3. The coax has become unbalanced and radiates. Also, the impedance of the dipole and coax changes due to the unbalanced currents and causes additional reflections. A balun is placed between the coax and dipole in order to cancel I_3. Various types of antenna elements require a balun [6]. Wideband baluns are more challenging to design, because the unwanted currents must be cancelled over the whole bandwidth.

4.2 ELEMENTS

This section lists a sampling of possible array elements. Wideband elements are emphasized, because a wideband array is often affiliated with a timed array.

4.2.1 Dipole Array

A transmission line feeds two arms as shown in Figure 4.5. Each arm is soldered to one of the transmission line conductors. Dipoles radiate best when the arms are half a wavelength long. A common configuration is the wire dipole shown in Figure 4.5a. This thin, half wavelength dipole has an approximate input impedance of 73Ω at its resonant frequency and is linearly polarized. A monopole is half of a dipole (Fig. 4.5b). One wire of the transmission line is fed through the hole in a ground plane and soldered to a quarter wavelength arm, while the other wire in the transmission line is soldered to the ground plane. Making the dipoles fatter (Fig. 4.5c) increases the bandwidth. A dipole with a length to diameter ratio of 5000 has a 3% bandwidth, while one with a length to diameter ratio of 260 has a 30% bandwidth [6].

A biconical antenna is a very broadband dipole variant in which the transmission line feeds the apex of two cones (Fig. 4.5d). Terminating the cones with hemispherical caps extends the bandwidth of the bicone even further (Fig. 4.5e). Other versions

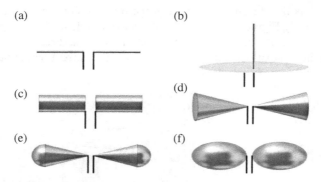

FIGURE 4.5 Dipole type antennas: (a) thin dipole, (b) monopole, (c) fat dipole, (d) bicone, (e) bicone with spherical caps, and (f) ellipsoidal.

FIGURE 4.6 Bent wire dipole element in the LWA1 array. Reprinted by permission of Ref. [9]; © 2013 IEEE.

of the dipole extend the bandwidth by increasing the thickness and decreasing sharp edges (Fig. 4.5f) [7]. Two-dimensional versions of these dipole-like antennas are broadband as well. For instance, the bow-tie antenna is a two-dimensional version of the bicone antenna [8]. Circular caps are often placed at the end of the bow-tie antennas as well. These antennas also come in monopole flavors too. Figure 4.6 shows part of the Long Wavelength Array Station 1 (LWA1) [9]. Each element is a wire-grid bowtie about 3 m long with the arms bent down by 45° in order to improve pattern uniformity over the sky. The feedpoint is 1.5 m above a 9 m² ground plane made from a 10 cm × 10 cm wire grid.

Figure 4.7 has two examples of dipole arrays. The molecular electronics for radar applications (MERAs) project was started by the US Air Force (USAF) in 1964 to develop an X-band all solid-state radar [10]. Its linearly polarized dipoles are shown in Figure 4.7a. The High Frequency Active Auroral Research Program (HAARP) is an HF array of 180 cross dipoles used to do upper atmospheric research in Alaska [11]. Crossed dipoles can be fed to produce polarization ranging from linear to circular.

4.2.2 Patch Array

A patch antenna originated as a resonant structure with a narrow bandwidth [13]. The simplest patch is etched from one side of a double-sided printed circuit board (PCB), while the bottom side serves as a ground plane. The patch may have various shapes with rectangular being the most popular. The patches are fed from the side or from

(a)

(b)

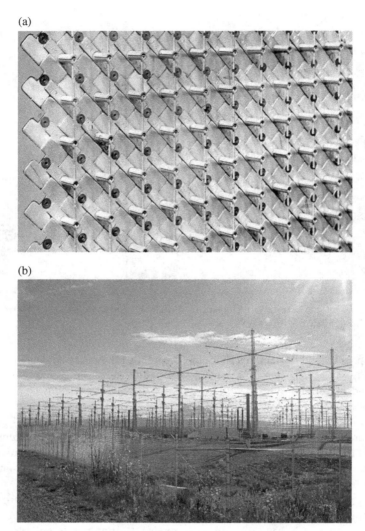

FIGURE 4.7 Dipole arrays. (a) Dipoles in MERA array. (Courtesy of the National Electronics Museum) [5] © 2013 Wiley. (b) Crossed dipoles in HAARP array (Courtesy Michael Kleiman, US Air Force) [12].

the bottom. These elements are easy to mass produce, can be made to conform to a surface, and are very rugged. Patches etched on a large PCB make a panel. Panels fit together with other panels to form very large arrays. Figure 4.8 is an 8×8 array of patch antennas with a corporate microstrip feed network.

The design parameters for a simple rectangular patch are the length (L), the width (W), the substrate height (h), and the substrate dielectric constant (ε_r). Two approaches to patch design are to either consider the patch as a resonant cavity or as a transmission

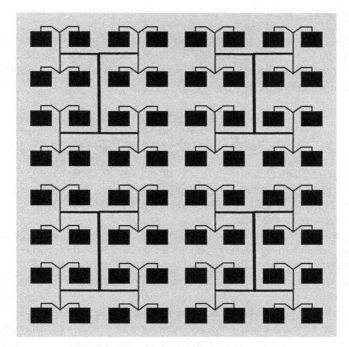

FIGURE 4.8 8×8 planar array of patches.

(a)   (b)

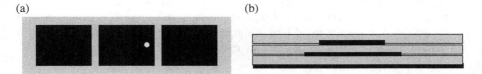

FIGURE 4.9 Broadband patch antennas: (a) coplanar and (b) stacked.

line [14]. The bandwidth of a patch can be increased by making the substrate thicker or decreasing the dielectric constant of the substrate [15]. Adding parasitic patches either next to the main patch or stacked on top of the patch significantly increases the beamwidth (Fig. 4.9) [16]. Increasing the patch area is not conducive to fitting the patch into the unit cell, so stacked patches are frequently used in arrays. A simple rectangular patch has only a few percent bandwidth, but stacked patches can be designed with over 20% bandwidth [17].

The patch polarization is determined by the location of the feed, shape of the patch, slots in the patch, or shorting vias [16]. There are two approaches to creating elliptically polarized patch antennas. First, feed the patch at a single point and change the shape of the patch or place slots inside the patch. Second, feed the patch at two points so the amplitude and phase difference create the desired polarization. In both cases, two modes are induced in the patch in order to form CP.

4.2.3 Spiral Array

A spiral antenna has one or more conducting wires or strips wrapped into a spiral [18]. Each arm of an N_{arm} spiral is fed with a current that has a phase increment of $360°/N_{arm}$ in order to radiate circular polarization. An Archimedean spiral (Fig. 4.10a) in the x–y plane is defined by

$$r = r_{in} + r_a \phi \qquad (4.8)$$

where ϕ is positive and not modulo 2π, r_{in} = distance from the center to the start of an arm, and r_a = growth rate.

The metal and dielectric in a self-complementary spiral have the same area and shape. Any self-complementary planar antenna has an input impedance of $Z_0/2 = 60\pi\Omega$ [19]. A spiral should have at least three turns [20]. Tight spirals tend to have better frequency-independent radiation patterns.

The logarithmic or equiangular spiral has all of its dimensions defined in linear proportion to a change in wavelength. Thus, the width of the conductor increases with distance from the feed (Fig. 4.10b). This type of spiral is considered "frequency independent" and has a very broad bandwidth. The equiangular spiral has smoother frequency performance over the bandwidth than an Archimedean spiral [20].

Spirals radiate right-/left-hand circularly polarized waves from the front while radiating left-/right-hand circularly polarized waves from the back. A ground plane forces radiation to the front size, but the reflection and coupling cause a large cross-polarization component. Placing a lossy cavity behind the spiral insures radiation from only one side but limits antenna efficiency to 50%. The spiral acts like a probe that excites a cavity. The cavity diameter is slightly larger than the spiral and its depth is at least one-quarter of a wavelength at the lowest frequency if no absorber is used. Not using absorber limits the spiral bandwidth to 2:1 [20]. Another alternative to the cavity backed spiral is the conical spiral (Fig. 4.10c), since it only radiates in the direction of the cone apex.

(a)   (b)   (c)

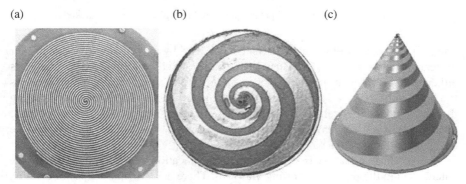

FIGURE 4.10 Examples of spiral antennas: (a) 2 arm Archimedes [5] © 2010, (b) two-arm logarithmic spiral. Reprinted by permission of Ref. [21]; © 2009 IEEE, and (c) conical log spiral [22] © 2010.

The circumference of the circle that contains the spiral approximately equals the maximum wavelength radiated by the spiral [23]. The lowest frequency in the bandwidth is determined by the outer radius of the spiral, while the high frequency is determined by the inner radius or smallest dimension.

$$\frac{c}{2\pi r_{\text{out}}} \leq f \leq \frac{c}{2\pi r_{\text{in}}} \qquad (4.9)$$

A spiral radiates from the active region defined by a disk with a one-wavelength circumference. At the lowest frequency in the bandwidth, the current reflects from the end of the arms. The reflected currents flow toward the spiral feed and radiate with the opposite sense of CP, thus degrading the polarization performance. Typically, the spiral diameter is made 20% larger than the upper limit of (4.9) in order to reduce reflections [20]. Another approach to reducing the reflections is to load the ends of the arms [20]. Placing absorber on the outer few rings preserves the spiral's CP and impedance at the expense of a gain loss of about 2 dB. Sometimes, RF chip resistors ($120-200\Omega$) are placed at the ends of the arms. Dielectric loading creates a reactive termination that slows the current and makes the arms look longer. This approach is less desirable, because it may introduce higher-order spiral radiation modes.

In an array, the spiral diameter must be less than the element spacing. For a linear array or planar array with a rectangular lattice, grating lobes limit the maximum spacing between elements to $d = \lambda_{\text{min}}/(1+\sin\theta_s)$. The lowest frequency limit occurs when the spiral diameter equals the element spacing or $\pi d = \lambda_{\text{max}} \Rightarrow d = \lambda_{\text{max}}/\pi$. The maximum possible bandwidth of a spiral array is the ratio of the maximum to the minimum wavelength.

$$\text{BW} = \frac{\lambda_{\text{max}}}{\lambda_{\text{min}}} = \begin{cases} \dfrac{\pi}{1+\sin\theta_s} & \text{for linear and rectangular lattice planar arrays} \\[4mm] \dfrac{1.15\pi}{1+\sin\theta_s} & \text{for hexagonal lattice planar arrays} \end{cases} \qquad (4.10)$$

This formula limits the bandwidth to about 3:1 for a spiral array.

The sinuous antenna is a broadband, planar or conical, rotationally symmetric structure comprising multiple zigzagging arms [24]. Like the spiral, the bandwidth is determined by the outer and inner diameters of the active antenna region. Unlike a spiral, the sinuous antenna has sections with arms that wind right-handed and sections with arms that wind left-handed that cause a polarization flip between neighboring bands and other multiband behavior when the antenna is fed in the difference mode [25]. It also needs a lossy cavity and is difficult to fit inside the unit cell. Figure 4.11 is an example of a four-arm sinuous antenna with seven bends. The 5 cm diameter antenna is etched on a 0.89 mm thick RT/Duroid 5880 substrate and has a flat 5-mm deep air-filled cavity.

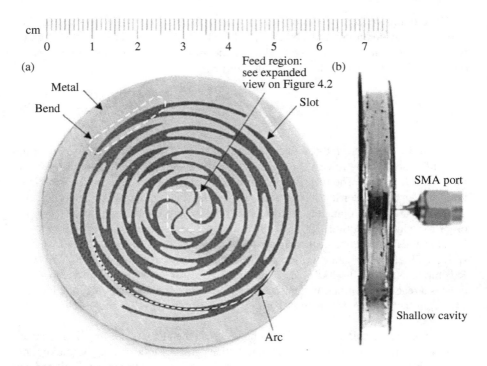

FIGURE 4.11   Photograph of the (a) front and (b) side views of a sinuous antenna. Reprinted by permission of Ref. [25]; © 2004 IEEE.

4.2.4   Helical Array

A helical antenna has the shape of a corkscrew [26]. The diameter, pitch, and number of turns in the helix control the polarization and directivity. If

$$N_{\text{helix}} \sqrt{s_{\text{helix}}^2 + (\pi D_{\text{helix}})^2} = \lambda \tag{4.11}$$

where D_{helix} = diameter of loop, s_{helix} = spacing between loops, and N_{helix} = number of loops then the helical antenna behaves like a monopole and operates in the normal mode. Normal meaning the maximum gain occurs perpendicular to the axis of the helix. The advantage of a normal mode helix is that it behaves like a monopole but is shorter at the same frequency.

An axial mode helix has its peak gain in the direction of its axis. It has an input impedance of [26]

$$Z_{\text{in}} = \frac{140\pi D_{\text{helix}}}{\lambda} \tag{4.12}$$

If a 50Ω input impedance is desired, then $D_{\text{helix}} = 0.1137\lambda$. The AR is [26]

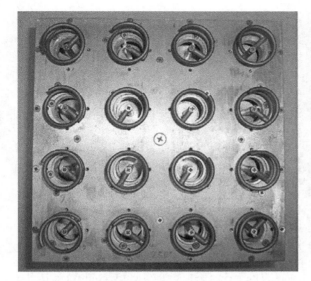

FIGURE 4.12 Geometry of a 4×4 helical array. Reprinted by permission of Ref. [27]; © 2010 IEEE.

$$AR = \frac{2N_{helix} + 1}{2N_{helix}} \qquad (4.13)$$

which is close to 1.0 (circular polarization). Figure 4.12 is a prototype 4×4 array of helical antennas with the center frequency of 4.0 GHz an element spacing of 45 mm [27].

4.2.5 Tapered Slot Antenna (TSA) Array

The tapered slot antenna (TSA) [28] is a very popular wideband element with the following characteristics [29]:

1. Bandwidth up to 12:1 [30].
2. $\cos\theta$ element pattern over all scan planes out to $\theta_s = 50° - 60°$.
3. No lossy materials needed to get wide bandwidth.
4. A balun can be easily incorporated into the design.
5. Beamwidth remains constant over the operating bandwidth.
6. Efficiency > 90%.
7. Linear polarized but can be arranged and fed to get circular polarization

Figure 4.13a is an exploded view of a 1–5 GHz TSA, and Figure 4.13b is a group of five of those elements that share common ground planes and dielectric substrates [31]. The element has two outer ground planes with flared slots that sandwich the

(a)

Groundplane layers

Strip
layer

Dielectric layers

(b)

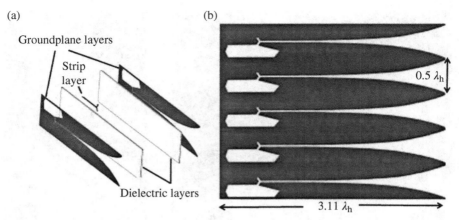

0.5 λ_h

3.11 λ_h

FIGURE 4.13 (a) Exploded view of the stripline notch element and (b) linear array of five elements. Reprinted by permission of Ref. [31]; © 2000 IEEE.

strip feed and two dielectric layers. The strip feed is electrically connected to the center conductor of a feeding coaxial line, while the outer conductor is connected to the ground plane.

A curved opening in the TSA reduces reflections and increases the bandwidth compared to a linear opening. Optimizing the exponential taper, length, and width of the slot as well as the feed can significantly improve the bandwidth. The feed design limits the high frequency, and the aperture size limits the low frequency of the bandwidth. Some general guidelines for the design of a TSA antenna array include the following [29]:

1. $d_x, d_y \approx 0.45\lambda$
2. Increasing the element length lowers the minimum frequency in the bandwidth.
3. Substrate:
   a. Its thickness should be about 1/10 the element width.
   b. Increasing ε_r lowers the minimum frequency in the bandwidth but may also lower the maximum frequency as well.
4. Increasing the opening rate of the aperture lowers the minimum frequency in the bandwidth but increases variations in the impedance over the bandwidth.
5. Increasing the cavity size increases the antenna resistance at lower frequencies.

Surrounding the TSA with vias that short the top and bottom metal surfaces on either side of the substrate eliminates unwanted resonances that limit the bandwidth [32].

This type of TSA requires a balun, because a balanced slot line feed connects to unbalanced microstrip. The balun must operate over a frequency range of greater than two octaves. An alternative to the TSA plus balun is the antipodal tapered slot antenna [33]. It has a tapered transition from microstrip, through parallel strip line, to a symmetric double side slot line and made a double-sided arrangement for the

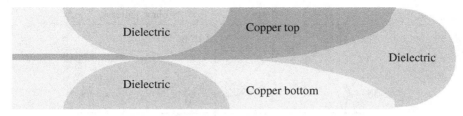

FIGURE 4.14   Antipodal TSA.

antenna (Fig. 4.14). This arrangement alleviates the difficulties of broadband coupling to slot line and the need for a balun.

The experimental 8×8 uniform array in Figure 4.13 scans to 50° in the E- and H-planes and had a VSWR < 2 at all scan angles from 0.75 to 1.3 GHz and 1.6 to 5 GHz [31]. Figure 4.15 shows the measured patterns at broadside. The co-pol and cross-pol array patterns in addition to a center element co-pol pattern are shown for both the E- and H-planes. Both measured and calculated active VSWRs are shown in Figure 4.16a for broadside, Figure 4.16b for a 50° scan in the E-plane, and Figure 4.16c for a 50° in the H-plane.

Figure 4.17 is a 324-element dual polarized TSA array for radio astronomy instrumentation [34]. The array has groups of four TSAs arranged in a cross-shaped structure. These TSA elements have a symmetrical main beam with an 87.5° beamwidth at 3 GHz and 44.2° beamwidth at 6 GHz. The measured maximum side/backlobe level is 10.3 dB below the main beam level. Grating lobes, not element bandwidth, limit the array operating frequency to below 5.4 GHz.

Figure 4.18 shows an array with body of revolution (BOR) elements that are dual polarized and broadband. This array has a 50Ω input impedance over 6–18 GHz and has a 45° conical grating lobe free scan volume [35]. The aluminum element screws into an aluminum ground plane making it easy to assemble, disassemble, and connect to active microwave modules. Since the array is solid aluminum and has no dielectric materials, conductor losses are negligible. At lower frequencies, machining the BOR elements from solid metal would make the array too heavy. Instead, the elements can be either metalized plastic or hollow metal.

4.2.6   Tightly Coupled Arrays

The current sheet antenna (CSA) array [36] (based on Wheelers' current sheet [37]) balances capacitive tightly coupled (close) dipoles with an inductive ground plane in order to get broad bandwidth. The planar ultrawideband modular antenna (PUMA) array (Fig. 4.19) is a current sheet antenna array with dipoles that have shorting vias connecting the arms and ground plane. The shorting vias are positioned to move resonances outside the system bandwidth [38]. Placing the shorting vias closer to the feed lines increases the element bandwidth. PUMA arrays have thick dielectric substrates that mechanically support the feed lines and shorting vias. The thick substrate also supports unwanted surface waves at certain scan angles. These surface waves are

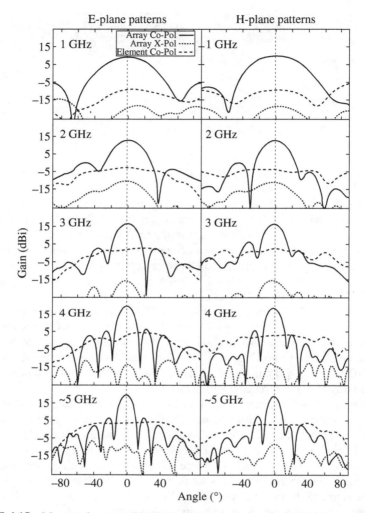

FIGURE 4.15 Measured cross-pol and co-pol patterns of an 8×8 planar array for broadside scan and the center element co-pol pattern. Reprinted by permission of Ref. [31]; © 2000 IEEE.

minimized by drilling holes in the substrate in order to lower the effective substrate permittivity.

Figure 4.20 shows three 8×8 PUMA modules (PUMAv1) that have a 3:1 bandwidth up to 21 GHz [39]. A second (PUMAv2) and third version (PUMAv3) have even wider bandwidths than the first version. Figure 4.21 has plots of the VSWR for all three versions of the PUMA arrays compared with a stacked patch. The PUMA array bandwidth dwarfs the bandwidth of the wideband stacked patch. Figure 4.21d compares the sizes of PUMAv2 and PUMAv3 with a penny.

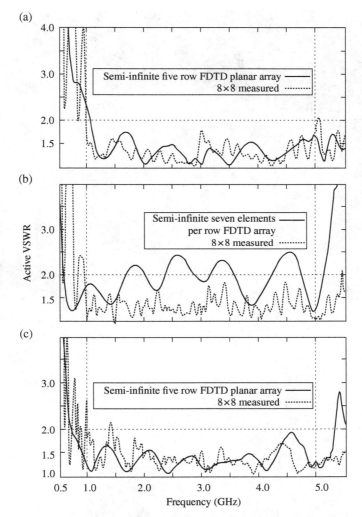

FIGURE 4.16 Measured versus calculated (FDTD) VSWR of planar arrays at (a) broadside scan, (b) E-plane 50 scan, and (c) H-plane 50 scan. Reprinted by permission of Ref. [31]; © 2000 IEEE.

Figure 4.22 is an example of a tightly coupled array of overlapping dipoles having a 5:1 bandwidth. The array is $\lambda_{lo}/10$ thick at the lowest frequency [40]. Since the dipole tips overlap, they are capacitively coupled to counteract the inductive effects of the ground plane.

Overlapped bowtie elements significantly improve the bandwidth of the tightly coupled dipole array through capacitive coupling and by adding a resistive frequency-selective surface (FSS) that suppress ground plane reflections.

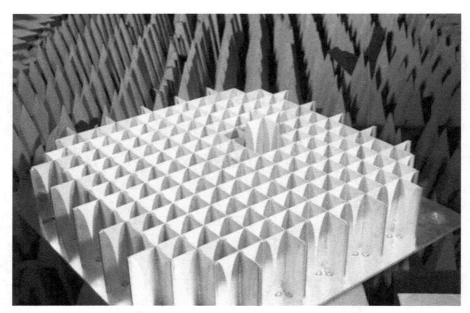

FIGURE 4.17 A 324-element dual-polarized TSA array. Reprinted by permission of Ref. [34]; © 2012 IEEE.

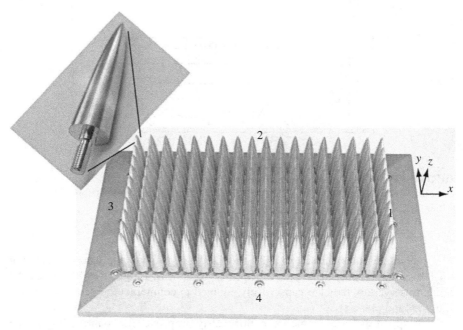

FIGURE 4.18 The BOR element array is 16×9 cm. Reprinted by permission of Ref. [35]; © 2007 IEEE.

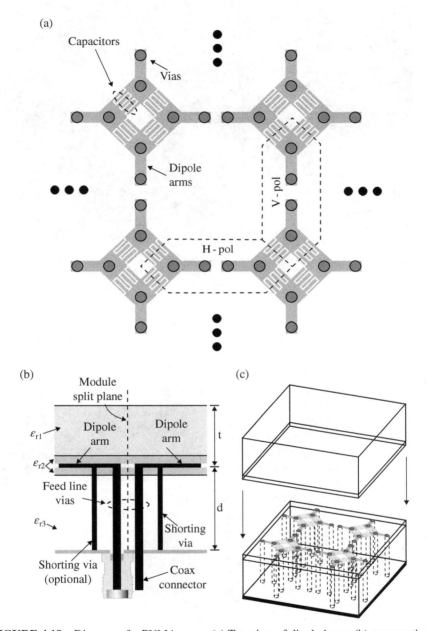

FIGURE 4.19 Diagram of a PUMA array: (a) Top view of dipole layer; (b) crosssectional view of a unit-cell, showing the location where a module split occurs; and (c) isometric view of a PUMA module with exploded dielectric cover layers. Reprinted by permission of Ref. [38]; © 2012 IEEE.

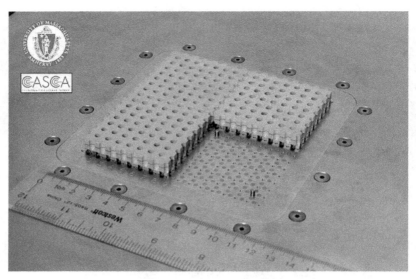

FIGURE 4.20 Three 8×8 PUMA modules. Reprinted by permission of Ref. [39]; © 2013 IEEE.

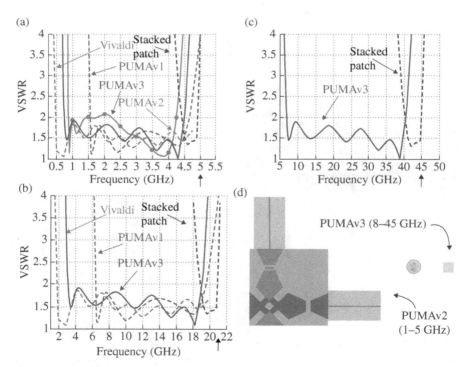

FIGURE 4.21 Broadside VSWR of three dual-polarized PUMA arrays: (a) 5 GHz maximum frequency. (b) 21 GHz maximum frequency. (c) 45 GHz maximum frequency. (d) Models of the PUMAv2 (left) and the PUMAv3 (right) compared to a penny. Reprinted by permission of Ref. [39]; © 2013 IEEE.

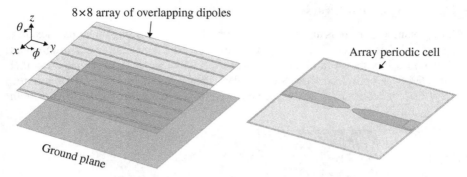

FIGURE 4.22 8×8 element array of overlapping dipoles above a ground plane. Reprinted by permission of Ref. [40]; © 2012 IEEE.

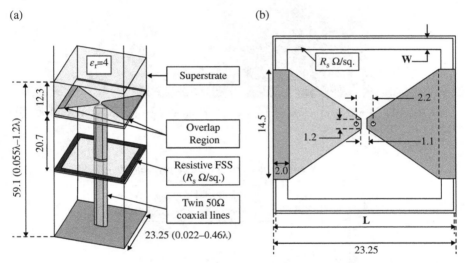

FIGURE 4.23 (a) Unit cell of a tightly coupled bowtie array with resistive FSS and superstrate and (b) top view of unit cell (dimensions in mm). Reprinted by permission of Ref. [41]; © 2012 IEEE.

The FSS makes the element lossy, though. Placing an appropriately designed superstrate mitigates those losses to only a little more than 1 dB yet significantly increases the bandwidth [41]. An element that is only $0.055\lambda_{lo}$ tall is shown in Figure 4.23. Simulations with this element in an infinite array show a 21:1 bandwidth and a radiation efficiency greater than 73%. The square ring-resistive elements in the FSS are polarization-insensitive, so it works with a dual-polarized crossed bowtie array.

4.2.7 Fragmented Arrays

A fragmented aperture is an optimized distribution of conducting metal patches on a substrate that radiates over a very broad bandwidth [29]. An example of a fragmented array unit cell is shown in Figure 4.24 [42]. A 21×21 element fragmented array that scans 60° was optimized from 8 to 12 GHz. Figure 4.25 is a plot of the measured VSWR of the whole array and of the center embedded element. The VSWRs of both

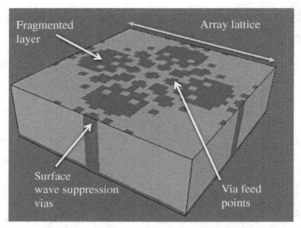

FIGURE 4.24 Unit cell of a fragmented array. Reprinted by permission of Rf. [42]; © 2011 IEEE.

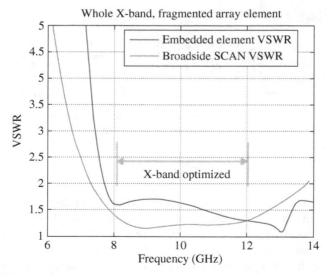

FIGURE 4.25 VSWR of fragmented element in planar array. Reprinted by permission of Ref. [42]; © 2011 IEEE.

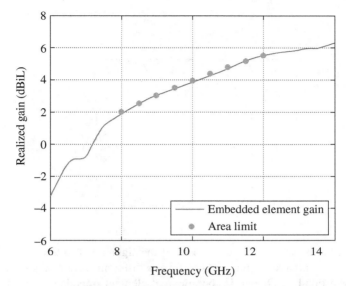

FIGURE 4.26 Gain of fragmented element in planar array. Reprinted by permission of Ref. [42]; © 2011 IEEE.

the embedded element and the array at broadside are below 2 over the desired frequency band. The measured realized gain of the embedded element is shown in Figure 4.26. The dots indicate the theoretical gain of the unit cell which is just a function of area.

4.3 MUTUAL COUPLING

When an isolated antenna is matched to its transmission line, then the antenna impedance (Z_a) equals the transmission line impedance (Z_c), and signals do not reflect from the antenna/transmission line interface, $\Gamma = 0$. An array complicates the element/ feed line matching because the elements transmit and receive to each other. Figure 4.27 is a diagram of a three-element transmit array. Even though the isolated elements are matched and the voltages fed to the elements (V_1^+, V_2^+, and V_3^+) are not reflected, there appear to be reflected signals (V_1^-, V_2^-, and V_3^-) due to the signals transmitted (coupled) from the other elements. These "reflected signals" give the appearance of an impedance mismatch. Since the reflection coefficient is no longer zero, the antenna impedance differs from the transmission line impedance by

$$Z_a = Z_c \frac{1 + \Gamma}{1 - \Gamma} \tag{4.14}$$

This mutual coupling of signals between elements is characterized by the mutual impedance.

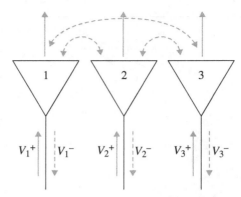

FIGURE 4.27   Mutual coupling in an array.

Steering the beam changes the phases of the signals fed to the elements. As the beam scans, the coupled signals add to form a different 'reflected signal" and produce a corresponding change in the apparent element impedance as a function of scan angle. The active impedance of an array element is the ratio of the voltage across the terminals of the element to the current flowing at those terminals when all the elements are excited. The active reflection coefficient of an array element is calculated using the active impedance.

Self-impedance is the impedance of the isolated antenna and is given by Ohm's law:

$$V_1 = Z_{11} I_1 \qquad (4.15)$$

The voltage at element 1 due to the current on element 2 is

$$V_1 = Z_{12} I_2 \qquad (4.16)$$

Z_{12} is the mutual impedance between elements 1 and 2. The total voltage at element 1 is the sum of the currents at the elements times the self- or mutual impedances.

$$V_1 = Z_{11} I_1 + Z_{12} I_2 + Z_{13} I_3 \qquad (4.17)$$

The impedance at element 1 is no longer Z_{11} when the other elements are present. The driving point or active impedance of the element is given by

$$Z_{a1} = \frac{V_1}{I_1} = Z_{11} + Z_{12} \frac{I_2}{I_1} + Z_{13} \frac{I_3}{I_1} \qquad (4.18)$$

The mutual impedance is a function of the distance between the elements, the polarization of the elements, the orientation of the elements, guided waves in a substrate connecting the elements, and the element patterns.

The mutual impedance between any two elements in the array is found by dividing the open-circuit voltage at one element by the current at the other element.

$$Z_{mn} = \frac{V_m}{I_n} \qquad (4.19)$$

Calculating all the impedances and placing them in an impedance matrix results in the matrix equation:

$$\begin{bmatrix} Z_{11} & Z_{12} & \cdots & Z_{1N} \\ Z_{21} & Z_{22} & \cdots & Z_{2N} \\ \vdots & \vdots & \ddots & \vdots \\ Z_{N1} & Z_{N2} & \cdots & Z_{NN} \end{bmatrix} \begin{bmatrix} I_1 \\ I_2 \\ \vdots \\ I_N \end{bmatrix} = \begin{bmatrix} V_1 \\ V_2 \\ \vdots \\ V_N \end{bmatrix} \qquad (4.20)$$

This $N \times N$ impedance matrix contains all the self- and mutual impedances of an N-element array. Reciprocity results in a symmetric of impedance matrix, $Z_{mn} = Z_{nm}$. The active impedance for element n in an N element array is

$$Z_{an} = \frac{I_1}{I_n} Z_{n1} + Z_{n2} \frac{I_2}{I_n} + \cdots + Z_{nN} \frac{I_N}{I_n} \qquad 4.21$$

Scan blindness results when the coupled waves destructively add with the transmitted wave to cause a large reflection coefficient [43–45]. In other words, the element steering phase matches the phase propagation of a surface wave on the aperture [44]. Complete scan blindness only occurs in infinite arrays, but large refection coefficients can occur in finite arrays [45]. Various techniques for breaking up the surface waves include baffles [46], metasurfaces [47], and vias.

Wide-band signals experience a time delay between the initial radiated pulse from one element and the re-radiated pulse from other elements. Depending on the pulse width, the time delay may be significant. This delay is at least equal to the travel time between the two elements.

The far field of a linear array when it transmits one delta function pulse is given by [48]

$$= \underbrace{\sum_{n=1}^{N} \delta\left[t + (n\sin\theta - n\sin\theta_s)d/c\right]}_{\text{Direct}} - \underbrace{\sum_{n=1}^{N} \sum_{\substack{m=1 \\ m \neq n}}^{N} \frac{G}{|n-m|} \delta\left[t + ()n\sin\theta - m\sin\theta_s - |n-m|d/c\right]}_{\text{Echoes}}$$

$$(4.22)$$

where $G \leq 1$ is a constant that is a function of element spacing and reflection coefficient. Figure 4.28 are graphs of the direct pulses and echoes from a four-element array with a 4 cm element spacing and $G = 1$. Although the amplitudes of the echoes are unrealistically high (G should be less than 1), the timing is accurate. Of course, there are no delta function pulses, so the pulses would be finite in length. These figures indicate that there could be overlap between pulses and echoes if the pulses have a long enough duration.

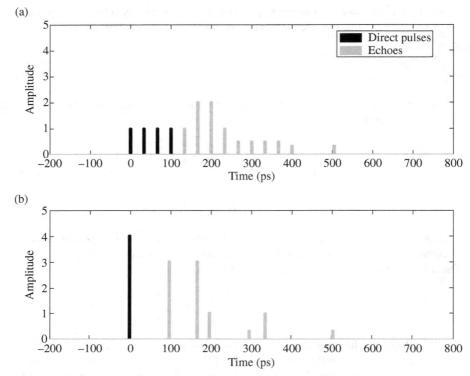

FIGURE 4.28 Graphs of the direct pulses and echoes for a 4-element array when $\theta_s = 15°$ and (a) $\theta = 0°$ (b). $\theta = 15°$. Reprinted by permission of Ref. [48]; © 2006 IEEE.

4.4 ELEMENT DISPERSION

The phase center of an element moves as a function of frequency and look angle. This movement is minor for small, two-dimensional antennas, but can be significant for other antennas, particularly for antennas that have height as well as area. Frequency-independent antennas radiate from different sections at different frequencies, so they are prone to dispersion. Small-scale features radiate high frequencies, while large-scale features radiate low frequencies. Since a wideband signal simultaneously has high and low frequencies, it radiates from multiple parts of the element. The path length from the high-frequency section is different than the path length from the low-frequency section, so the signal becomes dispersed. Figure 4.29 shows the normalized transmitted and received pulses by a pair of elliptical dipole antennas [49]. These small, planar antennas produce little pulse dispersion, because the phase center in the x–y plane stays in approximately the same place. By contrast, Figure 4.30 shows the normalized transmitted and received pulses by a pair of conical log-spiral antennas [49]. These frequency-independent nonplanar antennas exhibit substantial pulse distortion due to the different path lengths traveled

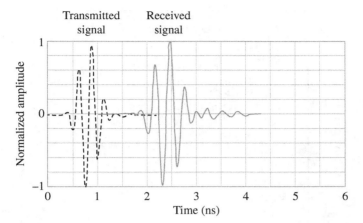

FIGURE 4.29 Pulse transmitted and received by two elliptical dipole antennas [49].

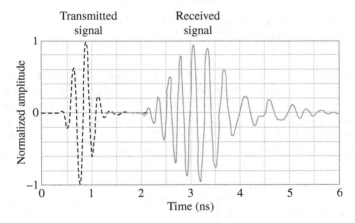

FIGURE 4.30 Pulse transmitted and received by two conical log spiral antennas [49].

by the different frequency components. In this case, the received pulse is over twice as long as the transmitted pulse. It exhibits a chirp in which the earlier part has a higher frequency content compared to the later part.

Antenna group delay is defined as follows:

$$\tau_g (f) = -\frac{d\psi (f)}{df} \tag{4.23}$$

where $\psi(f)$ is the frequency-dependent phase of the radiated signal. The average group delay is found by integrating over the bandwidth [21]:

$$\overline{\tau_g} = \frac{1}{f_{hi} - f_{lo}} \int_{f_{lo}}^{f_{hi}} \tau_g (f) df \tag{4.24}$$

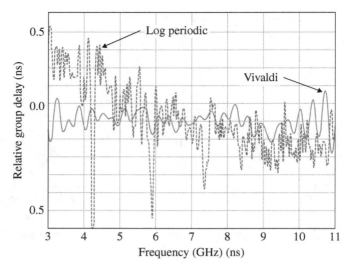

FIGURE 4.31 Relative group delay of a Vivaldi antenna and a log-periodic antenna. Reprinted by permission of Ref. [21]; © 2012 IEEE.

A nondistorted antenna has a constant group delay, that is, linear phase, in a relevant frequency range. Nonlinearities in the group delay indicate a resonance is present, which implies the antenna can store the energy. It results in ringing and oscillations of the antenna's transient response. Figure 4.31 compares the group delay of a Vivaldi antenna with a log-periodic antenna. The phase center of the Vivaldi is much more stable as a function of frequency than the phase center of the log-periodic antenna.

4.5 SCALED ARRAYS

The elements in an ultrawideband array are spaced far apart at high frequencies and close together at low frequencies. In order to maintain a relatively constant beamwidth across the operating bandwidth, a scaled array uses small wideband elements in the central region of a planar array, while increasingly larger elements with narrower bandwidth are used further from the center [50]. The design in Figure 4.32 has a central area of 4–18 GHz elements spaced $\lambda_{hi}/2$ apart. The elements in each ring around the central region increase in size and element spacing and maintain a constant 10° beamwidth in the principle planes.

Rather than using rings of elements with different bandwidths, the array can be divided into areas as shown in Figure 4.32. An efficient use of the aperture is possible by dividing it into smaller areas with each area designed to radiate over a portion of the bandwidth. A version of a wavelength-scaled array (WSA) was built using three TSA elements that operate in overlapping bands across from 1 to 8 GHz [51]. Figure 4.33 is a diagram of the array that has 64 elements with an 8:1 bandwidth,

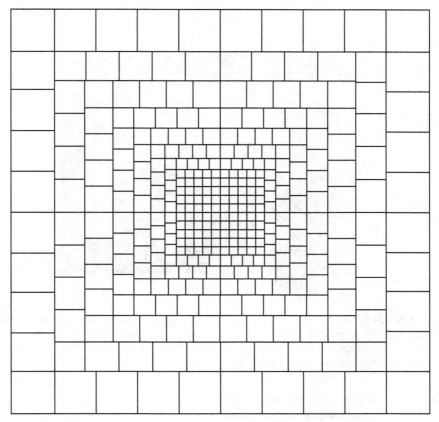

FIGURE 4.32 Scaled planar array with a 10° beamwidth in the principle planes from 4 to 18 GHz. Reprinted by permission of Ref. [50]; © 2006 IEEE.

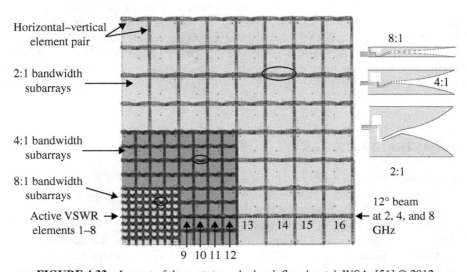

FIGURE 4.33 Layout of the prototype dual-pol. flared-notch WSA. [51] © 2012

(a)   (d)

(b)

(c)

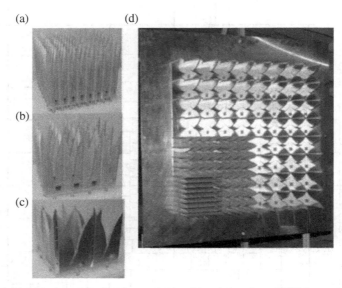

FIGURE 4.34 Modular construction and assembly of the planar WSA prototype: (a) 8×8 sub-array, 8:1 bandwidth (1–8 GHz); (b) 4×4 sub-array, 4:1 bandwidth (1–4 GHz); (c) 2×2 sub-array, 2:1 bandwidth (1–2 GHz); and (d) assembled dual-polarized WSA prototype. Reprinted by permission of Ref. [51]; © 2012 IEEE.

48 elements with a 4:1 bandwidth, and 48 elements with 2:1 bandwidth. This WSA replaces a 32×32 element ultrawideband array with the same footprint. The three levels of 2:1 scaling of element sizes gives this WSA a 6.4 reduction in the number of elements. A 32×32 element ultrawideband has a variable beam size over 1–8 GHz. The WSA is designed to have a 12° beamwidth at 2, 4, and 8. At 1–2 GHz, all elements are active. From 2 to 4 GHz, the outer 2:1 bandwidth elements are not active. From 4 to 8 GHz, only the 8:1 bandwidth elements are active. Figure 4.34a–c shows the three different TSAs that make up the dual-polarized planar WSA in Figure 4.34d. Figure 4.35 compares the measured and simulated pattern cuts in the E-plane. The measured results verify that the antenna patterns at 2, 4, and 8 GHz are nearly identical.

4.6   INTERLEAVED ARRAYS

An alternative to scaled arrays is interleaving elements that have different frequency coverage. Most interleaved arrays consist of two or three arrays operating at different frequencies [52–56]. The element spacing is uniform at all the operational bands, so avoiding grating lobes at the high frequencies is challenging. Unlike the scaled arrays, the arrays over the different frequency ranges do not have the same patterns.

Another approach to interleaving elements in an array creates two apertures operating over the same bandwidth from a single uniform array [57]. Figure 4.36 is

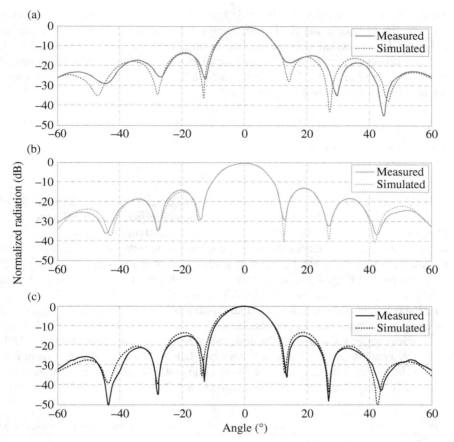

FIGURE 4.35 E-plane radiation patterns at the three breakpoints: (a) 2 GHz, (b) 4 GHz, and (c) 8 GHz, showing agreement between simulations and measurement. Reprinted by permission of [51]; © 2012 IEEE.

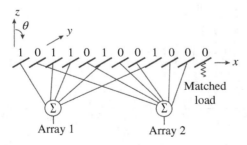

FIGURE 4.36 Diagram of an interleaved array of dipoles. Reprinted by permission of Ref. [57]; © 2005 IEEE.

a diagram of an interleaved dipole array. The output from a dipole connects to either array 1, array 2, or a matched load. Arrays 1 and 2 are different beamforming networks. Elements that serve neither beamformer are connected to a matched load, and that dipole is inactive. Consider the case when one array has an element connected to the feed network while the other array has it disconnected, so the amplitude weights for the first array are represented by

$$\mathbf{w} = [w_1, w_2, \ldots, w_N, 1 - w_N, \ldots, 1 - w_2, 1 - w_1] \qquad (4.25)$$

and for the second array by

$$\mathbf{w}' = 1 - \mathbf{w} \qquad (4.26)$$

A thinned sum array has the highest density of elements turned on in the center of the aperture and much fewer at the edges. A thinned difference array has few elements turned on in the center and at the edges. A genetic algorithm found the optimum thinning arrangement that results in the minimum maximum sidelobe level for both the sum and difference array factors of a 120-element linear array of dipoles [57]. The sum pattern (Fig. 4.37) has a 45% aperture efficiency with 13.1 dB peak sum sidelobes. The difference pattern (Fig. 4.38) has a 55% efficiency with 13.3 dB. The entire aperture is 100% efficient, so optimum use is made of the available space and no element is terminated in a matched load.

Interleaving was extended to the design of a dual-polarized spiral array [58]. Half of the elements were RHCP and the other half LHCP. The array operates from 2 to 6 GHz and has an element spacing of 38 mm. Figure 4.39 shows a portion of the array where a "0" designates an LHCP element and a "1" designates an RHCP element.

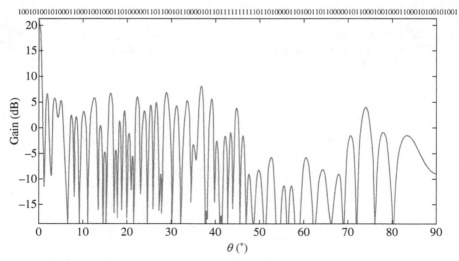

FIGURE 4.37 Sum array factor for a 120-element aperture that has sum and difference arrays interleaved. Reprinted by permission of Ref. [57]; © 2005 IEEE.

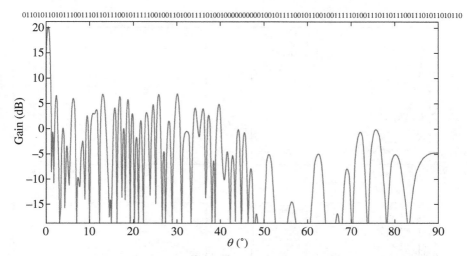

01101011010111001110110111001011111001001101001111010010000000000100101111001011001001111101001110110111001110101010110

FIGURE 4.38 Difference array factor for a 120-element aperture that has sum and difference arrays interleaved. Reprinted by permission of Ref. [57]; © 2005 IEEE.

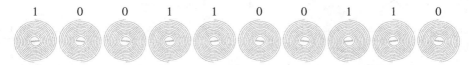

| 1 | 0 | 0 | 1 | 1 | 0 | 0 | 1 | 1 | 0 |

FIGURE 4.39 Example of a dual polarization spiral array, 1001100110. Reprinted by permission of Ref. [58]; © 2010 IEEE.

The array design goals include a ±30° scan with an AR less than 3 dB, a VSWR less than 2 and a relative sidelobe level (RSLL) under 10 dB. The array has 80 spiral elements that are modeled using the method of moments. Step 1 optimizes the isolated spiral element to get the best AR and VSWR. Next, the optimized element is placed in the array and the entire array is optimized over the scan range. The resulting representation is

[1010011110110100101010110111000001100011001110011111000100101 0101101001000011010]

Figure 4.40a is a plot of the AR of the array when steered to 0, 10, 20, and 30°. The AR is under 3 dB at frequencies above 4 GHz for all scan angles. Figure 4.40b shows the VSWR of element 20 (worst case) when matched to 250 Ω for the four scan angles. There is a small resonance at 3.25 GHz when the beam is steered to 20° or 30°. The VSWR is less than 2 between 2.75 and 10 GHz (except around 3.25 GHz, but this exception only occurs for element 20). The array bandwidth is limited by the AR and not the VSWR. Figure 4.40c is a plot of the array pattern at broadside for 3 GHz, and in Figure 4.40d the pattern is steered to 30° at 5 GHz. Assuming the bandwidth is defined for an RSLL less than −10 dB, an AR less than 3 dB and a VSWR

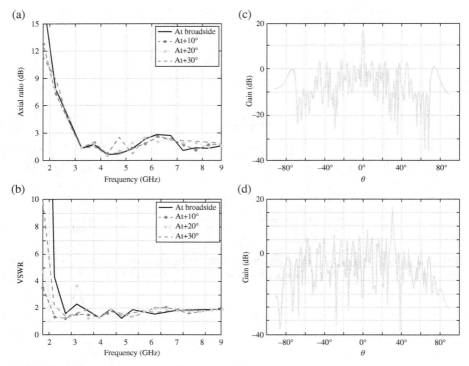

FIGURE 4.40 Characteristics of an 80-spiral antenna array. These spirals are standard 2-arms center-fed Archimedean spirals, they have five turns, for a diameter of 34 mm. The distance center to center in the array is 38 mm. (a) Axial ratio. (b) VSWR at 250 Ω. Samples of the radiation patterns for a standard array. (c) At 3 GHz and at broadside. (d) At 5 GHz and steered at 30° from broadside. Reprinted by permission of Ref. [58]; © 2010 IEEE.

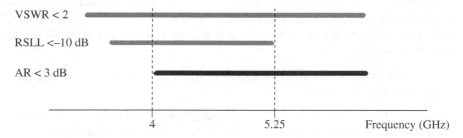

FIGURE 4.41 Bandwidth of a simple interleaved array of spiral with alternating polarization, following the three criteria. Reprinted by permission of Ref. [58]; © 2010 IEEE.

less than 2 oversteering angles from 0 to 30°. This array has an approximate bandwidth of 4–5.25 GHz, or 27% as shown in Figure 4.41. This bandwidth applies to both polarizations. The AR sets the lowest frequency in the bandwidth, while the RSLL sets the highest frequency. The VSWR bandwidth extends above and below these limits. Decreasing the element spacing in the array increases the upper frequency limit. Unfortunately, decreasing the spacing shrinks the size of the spiral, which in turn reduces the spiral bandwidth.

REFERENCES

[1] *IEEE Standard for Definitions of Terms for Antennas*, IEEE Standard 145™-2013.

[2] J. Huang, "A technique for an array to generate circular polarization with linearly polarized elements," *IEEE Trans. Antennas Propag.*, vol. 34, pp. 1113–1124, 1986.

[3] A. E. Tan *et al.*, "Time domain characterization of circularly polarized ultrawideband array," *IEEE Trans. Antennas Propag.*, vol. 58, pp. 3524–3531, 2010.

[4] W. Wasylkiwskyj and W. K. Kahn, "Element patterns and active reflection coefficient in uniform phased arrays," *IEEE Trans. Antennas Propag.*, vol. 22, pp. 207–212, 1974.

[5] R. L. Haupt, *Antenna Arrays: A Computational Approach*. Hoboken, NJ: John Wiley & Sons, Inc., 2010.

[6] C. A. Balanis, *Antenna Theory: Analysis and Design* (3rd ed.). Hoboken, NJ: Wiley-Interscience, 2005.

[7] B. Allen, *Ultra-Wideband: Antennas and Propagation for Communications, Radar and Imaging*. Hoboken, NJ: John Wiley & Sons, Inc., 2007.

[8] G. H. Brown and J. O. M. Woodward, "Experimentally determined radiation characteristics of conical and triangular antennas," *RCA Rev.*, vol. 13, pp. 425–452, 1952.

[9] S. W. Ellingson *et al.*, "Design and commissioning of the LWA1 radio telescope," *IEEE Trans. Antennas Propag.*, vol. 61, pp. 776–783, 2013.

[10] D. N. McQuiddy Jr. *et al.*, "Transmit/receive module technology for X-band active array radar," *Proc. IEEE*, vol. 79, pp. 308–341, 1991.

[11] *High-frequency active Auroral research program (HAARP)* [online]. Available: http://www.haarp.alaska.edu/haarp/index.html. Accessed December 17, 2014.

[12] J. Ohab, *Armed with Science. The Official U.S. Defense Department Science Blog, HAARP's antenna array: the test kitchen in the sky* [online]. Available: http://science.dodlive.mil/2010/02/23/haarps-antenna-array-the-kitchen-in-the-sky/. Accessed February 23, 2010.

[13] R. E. Munson, "Conformal microstrip antennas and microstrip phased arrays," *IEEE Trans. Antennas Propag.*, vol. 22, pp. 74–78, 1974.

[14] D. R. Jackson and N. G. Alexopoulos, "Simple approximate formulas for input resistance, bandwidth, and efficiency of a resonant rectangular patch," *IEEE Trans. Antennas Propag.*, vol. 39, pp. 407–410, 1991.

[15] M. Kara, "Formulas for the computation of the physical properties of rectangular microstrip antenna elements with various substrate thicknesses," *Microw. Opt. Technol. Lett.*, vol. 12, pp. 234–239, 1996.

[16] R. Garg *et al.*, *Microstrip Antenna Design Handbook*. Norwood, MA: Artech House, 2001.

[17] W. L. Stutzman and G. A. Thiele, *Antenna Theory and Design*. Hoboken, NJ: John Wiley & Sons, Inc., 2013.

[18] J. Dyson, "The equiangular spiral antenna," *IRE Trans. Antennas Propag.*, vol. 7, pp. 181–187, 1959.

[19] W. Curtis, "Spiral antennas," *IRE Trans. Antennas Propag.*, vol. 8, pp. 298–306, 1960.

[20] T. Lam *et al.*, "Spiral antenna design considerations," *Microw. J.*, vol. 56, pp. 84–94, 2013.

[21] W. Wiesbeck *et al.*, "Basic properties and design principles of UWB antennas," *Proc. IEEE*, vol. 97, pp. 372–385, 2009.

[22] H. Nakano *et al.*, "Spiral antenna radiating a conical beam of circular polarization," International Conference on Electromagnetics in Advanced Applications (ICEAA), September 2010, pp. 109–112.

[23] J. Kaiser, "The Archimedean two-wire spiral antenna," *IRE Trans. Antennas Propag.*, vol. 8, pp. 312–323, 1960.

[24] R. H. DuHamel, "Dual polarized sinuous antennas," U.S. Patent 4 658 262, April 14, 1987.

[25] M. C. Buck and D. S. Filipovic, "Split-beam mode four-arm slot sinuous antenna," *IEEE Antennas Wireless Propag. Lett.*, vol. 3, pp. 83–86, 2004.

[26] J. D. Kraus and R. J. Marhefka, *Antennas for All Applications* (3rd ed.). New York: McGraw-Hill, 2002.

[27] X. Li *et al.*, "16-Element single-layer rectangular radial line helical array antenna for high-power applications," *IEEE Antennas Wireless Propag. Lett.*, vol. 9, pp. 708–711, 2010.

[28] P. J. Gibson, "The Vivaldi aerial," Proceedings of the 9th European Microwave Conference, Brighton, England, September 1979, pp. 103–105.

[29] W. F. Croswell *et al.*, "Wideband arrays," in *Modern Antenna Handbook*, C. A. Balanis, Ed. Hoboken, NJ: John Wiley & Sons, Inc., 2008, pp. 581–630.

[30] R. Kindt and R. Pickles, "Ultrawideband all-metal flared-notch array radiator," *IEEE Trans. Antennas Propag.*, vol. 58, pp. 3568–3575, 2010.

[31] M. Kragalott *et al.*, "Design of a 5:1 bandwidth stripline notch array from FDTD analysis," *IEEE Trans. Antennas Propag.*, vol. 48, pp. 1733–1741, 2000.

[32] H. Holter *et al.*, "Elimination of impedance anomalies in single- and dual-polarized endfire tapered slot phased arrays," *IEEE Trans. Antennas Propag.*, vol. 48, pp. 122–124, 2000.

[33] E. Gazit, "Improved design of the Vivaldi antenna," *IEE Proc. H, Microw Antennas Propag*, vol. 135, pp. 89–92, 1988.

[34] E. W. Reid *et al.*, "A 324-element Vivaldi antenna array for radio astronomy instrumentation," *IEEE Trans. Instrum.*, vol. 61, pp. 241–250, 2012.

[35] H. Holter, "Dual-polarized broadband array antenna with BOR-elements, mechanical design and measurements," *IEEE Trans. Antennas Propag.*, vol. 55, pp. 305–312, 2007.

[36] B. Munk *et al.*, "A low-profile broadband phased array antenna," IEEE Antennas and Propagation Society International Symposium, Columbus, OH, USA, June 2003, pp. 448–451.

[37] H. A. Wheeler, "Simple relations derived from a phased-array antenna made of an infinite current sheet," *IEEE Trans. Antennas Propag.*, vol. 13, pp. 506–514, 1965.

[38] S. S. Holland and M. N. Vouvakis, "The planar ultrawideband modular antenna (PUMA) array," *IEEE Trans. Antennas Propag.*, vol. 60, pp. 130–140, 2012.

[39] J. T. Logan *et al.*, "A review of planar ultrawideband modular antenna (PUMA) arrays," URSI International Symposium on Electromagnetic Theory (EMTS), Hiroshima, Japan, 2013, pp. 868–871.

[40] I. Tzanidis *et al.*, "Characteristic excitation taper for ultrawideband tightly coupled antenna arrays," *IEEE Trans. Antennas Propag.*, vol. 60, pp.1777–1784, 2012.

[41] W. F. Moulder *et al.*, "Superstrate-enhanced ultrawideband tightly coupled array with resistive FSS," *IEEE Trans. Antennas Propag.*, vol. 60, pp. 4166–4172, 2012.

[42] J. G. Maloney *et al.*, "Wide scan, integrated printed circuit board, fragmented aperture array antennas," IEEE International Symposium on Antennas and Propagation (APSURSI), July 2011, pp. 1965–1968.

[43] L. Lechtreck, "Effects of coupling accumulation in antenna arrays," *IEEE Trans. Antennas Propag.*, vol. 16, pp. 31–37, 1968.

[44] D. Pozar and D. Schaubert, "Scan blindness in infinite phased arrays of printed dipoles," *IEEE Trans. Antennas Propag.*, vol. 32, pp. 602–610, 1984.

[45] G. H. Knittel *et al.*, "Element pattern nulls in phased arrays and their relation to guided waves," *Proc. IEEE*, vol. 56, pp. 1822–1836, 1968.

[46] S. Edelberg and A. Oliner, "Mutual coupling effects in large antenna arrays II: compensation effects," *IRE Trans. Antennas Propag.*, vol. 8, pp. 360–367, 1960.

[47] Z. Lijun *et al.*, "Scan blindness free phased array design using PBG materials," *IEEE Trans. Antennas Propag.*, vol. 52, pp. 2000–2007, 2004.

[48] M. Ciattaglia and G. Marrocco, "Investigation on antenna coupling in pulsed arrays," *IEEE Trans. Antennas Propag.*, vol. 54, pp. 835–843, 2006.

[49] H. G. Schantz, "Dispersion and UWB antennas," Joint UWBST & IWUWBS. 2004 International Workshop on Ultra Wideband Systems, Kyoto, Japan, May 18–21, 2004, pp. 161–165.

[50] B. Cantrell *et al.*, "Wideband array antenna concept," *IEEE Aerosp. Electron. Syst. Mag.*, vol. 21, pp. 9–12, 2006.

[51] R. W. Kindt, "Prototype design of a modular ultrawideband wavelength-scaled array of flared notches," *IEEE Trans. Antennas Propag.*, vol. 60, pp. 1320–1328, 2012.

[52] J. Hsiao, "Analysis of interleaved arrays of waveguide elements," *IEEE Trans. Antennas Propag.*, vol. 19, pp. 729–735, 1971.

[53] J. Boyns and J. Provencher, "Experimental results of a multifrequency array antenna," *IEEE Trans. Antennas Propag.*, vol. 20, pp. 106–107, 1972.

[54] K. Lee *et al.*, "A dual band phased array using interleaved waveguides and dipoles printed on high dielectric substrate," IEEE AP-S Symposium, Boston, MA, June 1984, pp. 886–889.

[55] R. Chu *et al.*, "Multiband phased-array antenna with interleaved tapered-elements and waveguide radiators," IEEE AP-S Symposium, Baltimore, MD, December 1996, pp. 21–26.

[56] K. Lee *et al.*, "Dual-band, dual-polarization, interleaved cross-dipole and cavity-backed disc elements phased array antenna," IEEE AP-S Symposium, Montreal, Quebec, Canada, July 1997, pp. 694–697.

[57] R. L. Haupt, "Interleaved thinned linear arrays," *IEEE Trans. Antennas Propag.*, vol. 53, pp. 2858–2864, 2005.

[58] R. Guinvarćh and R. L. Haupt, "Dual polarization interleaved spiral antenna phased array with an octave bandwidth," *IEEE Trans. Antennas Propag.*, vol. 58, pp. 1–7, 2010.

5

ARRAY BEAMFORMING

An array feed or beamforming network combines the signals from all the elements to form a receive beam or, conversely, distributes a transmitted signal to the elements in the array to create a transmit beam. This chapter presents an assortment of analog and digital beamforming networks for creating and manipulating beams. A corporate feed has power splitters/combiners that guide the signals to/from the elements. One port in a receive array collects the signals from all the elements to form one beam. A multibeam array needs a corporate feed for each beam. The Blass matrix and the Butler matrix use transmission line to generate simultaneous multiple beams. Bootlace lenses create simultaneous multiple beam approaches without using transmission lines. Finally, the most versatile and advanced beamforming network is the digital beamformer which places an RF analog to digital converter (ADC) at each element and sends the bit representation of the signal at each element to the computer. Beams are formed in software rather than in RF hardware.

5.1 PCB TRANSMISSION LINES

Most feed networks have transmission lines, such as coaxial cables, stripline, and microstrip, and passive devices, such as couplers and power splitters or active components, such as amplifiers and receivers. Impedance matching at the connections ensures efficient power transfer through the feed network. In order to simplify matching, most devices and transmission lines have a characteristic impedance of 50Ω.

Timed Arrays: Wideband and Time Varying Antenna Arrays, First Edition. Randy L. Haupt.
© 2015 John Wiley & Sons, Inc. Published 2015 by John Wiley & Sons, Inc.

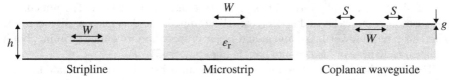

FIGURE 5.1 Three types of planar transmission lines.

Consequently, most antennas are designed to have a 50Ω input impedance as well. Figure 5.1 shows the three types of planar transmission lines that are presented next.

Stripline has a thin (g thick), narrow conductor inside a dielectric substrate between two large ground planes. Figure 5.1 shows a cross section of a stripline of width W suspended in a substrate half way between two ground planes separated by h. The characteristic impedance is a function of W, h, g, and ε_r and to within an accuracy of 6% is given by [1]

$$Z_c = \frac{60}{\sqrt{\varepsilon_r}} \ln\left(\frac{1.9h}{0.8W + g}\right) \tag{5.1}$$

The time delay over a symmetric stripline is

$$T_d = 33.5\sqrt{\varepsilon_r} \quad \text{ps/cm} \tag{5.2}$$

For an FR-4 substrate ($\varepsilon_r = 4$), when $2 \leq W/h \leq 2.2$, $Z_c \approx 50\Omega$ and $T_d = 70$ ps/cm. Stripline has good immunity to crosstalk and is resistant to electromagnetic interference.

Microstrip has a conducting strip lying on top of the substrate with a conducting ground plane beneath (Fig. 5.1). Microstrip is cheap, light weight, and compact. Active components are easily mounted on top of the board. The substrate has a thin copper clad on both sides. One side serves as the ground plane, while the circuit design is etched or routed from the other side. Since part of the electromagnetic wave carried by a microstrip line exists in the substrate and part in the air, the electric field surrounding the microstrip exists in two different media, so it has an effective dielectric constant given by [2]

$$\varepsilon_{eff} = \frac{\varepsilon_r + 1}{2} + \frac{\varepsilon_r - 1}{2} \frac{1}{\sqrt{1 + 12h/W}} \tag{5.3}$$

where, h = substrate thickness and W = width of microstrip line

The characteristic impedance of a microstrip line to within an accuracy of 5% is [1]

$$Z_c = \frac{87}{\sqrt{\varepsilon_r + 1.41}} \ln\left(\frac{5.98h}{0.8W + g}\right) \tag{5.4}$$

The time delay over a microstrip line is

$$T_d = 3.3\sqrt{\varepsilon_{eff}} \quad \text{ns/cm} \tag{5.5}$$

An FR-4 substrate with $W/h = 2$ has $Z_c \simeq 50\,\Omega$ and $T_d = 58\,\text{ps/cm}$. The impedance increases as h increases and ε_r decreases. The impedance decreases as W increases. Microstrip has a lower loss-tangent and faster propagation time than stripline, but it radiates, is dispersive, and cannot handle high power [3].

The transmission lines in the feed networks must bend in order to guide the signals to/from the elements. A sharp $90°$ bend produces large reflections. Mitering the bend or making it into an arc mitigates the reflections. If the bend is an arc of radius at least three times the strip width, then reflections are minimal. This large bend takes up a lot of real estate compared to the $90°$ bend, though. Douville and James experimentally modeled the optimum miter for a $50\,\Omega$ microstrip line over a wide range of microstrip geometries [4].

Coplanar waveguide (Fig. 5.1) is a trace on top of a dielectric substrate with two ground planes on either side separated by a distance, s. A variant called grounded coplanar waveguide has an additional ground plane at the bottom of the substrate. Active devices are easy to mount on top of coplanar waveguide, and it functions well at very high frequencies. In addition, connecting series and shunt circuit elements is possible without using metal filled via holes as required for microstrip and stripline.

The characteristic impedance of coplanar waveguide is given by [5]

$$Z_c = \frac{30\pi K'(k_1)}{\sqrt{\varepsilon_{\text{eff}}}\, K(k_1)} \tag{5.6}$$

$$\varepsilon_{\text{eff}} = 1 + \frac{(\varepsilon_r - 1)\,K(k_2)\,K'(k_1)}{2K'(k_2)\,K(k_1)} \tag{5.7}$$

where

$$k_1 = \frac{W}{W + 2s}$$

$$k_2 = \frac{\sinh(\pi W/4h)}{\sinh[\pi(W + 2s)/4h]}$$

$K(\bullet) = $ elliptical integral of the first kind

$$\frac{K(k)}{K'(k)} = \begin{cases} \dfrac{\pi}{\ln\left[2\dfrac{1+(1-k^2)^{0.25}}{1-(1-k^2)^{0.25}}\right]} & 0 \le k \le \dfrac{1}{\sqrt{2}} \\[4ex] \dfrac{\ln\left[2\dfrac{1+\sqrt{k}}{1-\sqrt{k}}\right]}{\pi} & \dfrac{1}{\sqrt{2}} \le k \le 1 \end{cases}$$

The impedance decreases as h increases.

CPW backed by an infinite ground plane has the following impedance [5]:

$$Z_c = \frac{60\pi}{\sqrt{\varepsilon_{\text{eff}}}\left[\dfrac{K(k_1)}{K'(k_1)} + \dfrac{K(k_3)}{K'(k_3)}\right]} \tag{5.8}$$

$$\varepsilon_{\text{eff}} = 1 + \frac{(\varepsilon_r - 1)(K(k_3)/K'(k_3))}{(K(k_1)/K'(k_1)) + (K(k_3)/K'(k_3))} \tag{5.9}$$

where

$$k_3 = \frac{\tanh(\pi W/4h)}{\tanh[\pi(W + 2s)/4h]}$$

The impedance increases as h increases. For best performance, $W + 2s < \lambda/2$ and the ground should extend $5(W + 2s)$ on either side of the trace. It has a lower loss tangent than microstrip but higher skin effect losses [2]. An FR-4 substrate with $W = 2.9\,\text{mm}$ and $s = 1.5\,\text{mm}$ has $Z_c \approx 50\,\Omega$ and $T_d = 57\,\text{ps/cm}$.

5.2   S-PARAMETERS

Scattering or S-parameters (S_{mn}) of an N port device are the reflection and transmission coefficients between the output port, m, and the input port, n. The voltage signal at port n is

$$V_n^- = S_{n1}V_1^+ + \cdots + S_{nn}V_n^+ + \cdots + S_{nN}V_N^+ \tag{5.10}$$

where

$$S_{nm} = \left.\frac{V_n^-}{V_m^+}\right|_{V_p^+ = 0 \ \text{for } p \neq m} \tag{5.11}$$

Writing similar equations for the other ports results in a matrix equation

$$\begin{bmatrix} S_{11} & \cdots & S_{1N} \\ \vdots & \ddots & \vdots \\ S_{N1} & \cdots & S_{NN} \end{bmatrix} \begin{bmatrix} V_1^+ \\ \vdots \\ V_N^+ \end{bmatrix} = \begin{bmatrix} V_1^- \\ \vdots \\ V_N^- \end{bmatrix} \tag{5.12}$$

where the $N \times N$ matrix is the S-parameter matrix. A loss-free network implies that the sum of the input signals equals the sum of the output signals, $\sum_{n=1}^{N} V_n^+ = \sum_{n=1}^{N} V_n^-$, while lossy passive network implies $\sum_{n=1}^{N} V_n^+ > \sum_{n=1}^{N} V_n^-$. Active networks may have transmission coefficients greater than one. The S-parameter matrix of a reciprocal network equals its transpose. Ferrites and amplifiers in a network make it nonreciprocal.

5.3   MATCHING CIRCUITS

Matching circuits reduce reflections between two transmission lines with different impedances (Z_1 and Z_3). A quarter-wave transformer is a section of transmission line $\lambda_c/4$ long at the center frequency with an impedance given by

$$Z_2 = \sqrt{Z_1 Z_3} \qquad (5.13)$$

The quarter wave transformer bandwidth is [2]

$$BW = 2 - \frac{4}{\pi}\cos^{-1}\left(\frac{2\Gamma_m Z_2}{|Z_3 - Z_1|\sqrt{1-\Gamma_m^2}}\right) \qquad (5.14)$$

where Γ_m is the maximum reflection coefficient. The bandwidth decreases as the difference between Z_1 and Z_3 increases. A broadband match has additional quarter-wave sections added between the two original impedances. The impedances are found through a number of synthesis approaches like binomial, Chebyshev, and so on [2]. Extending the bandwidth in this manner takes additional space, which is usually at a premium.

5.4   CORPORATE FEEDS

A corporate or parallel feed distributes transmit and/or receive signals in the array using couplers and power dividers/combiners. Three commonly used power dividers are resistive, quadrature hybrid, and Wilkinson. Their characteristics are summarized in Table 5.1. Example diagrams appear in Figure 5.2.

Resistive power dividers are small and broadband but have resistive power loss and the ports are not isolated [6]. A three-port resistive divider has 3 dB of resistive loss and a 3 dB power split. A signal at port 2 or 3 in Figure 5.2a is 6 dB less than the signal entering port 1. The 3 dB quadrature hybrid coupler is a four-port device with two inputs and two outputs (Fig. 5.2b). An input signal splits equally between the two outputs, with one of the outputs having a 90° phase shift. A Wilkinson power divider splits the input signal into two parts (Fig. 5.2c) [7]. Since this device is reciprocal, it serves as a power combiner at the same time. It is lossless when all the output ports

TABLE 5.1   Power Divider Characteristics[a]

	Resistive	Wilkinson	Quadrature hybrid
Frequency range	DC –10's GHz	100's MHz –10's GHz	100's MHz –10's GHz
Insertion loss	6 dB for two outputs	10 log(N) for N outputs	3 dB
Isolation	6 dB	20 dB	20 dB
Output phase shift	0	0	90°

[a] Ref. [6].

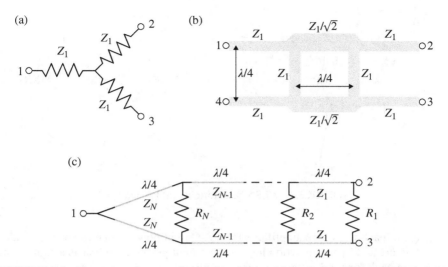

FIGURE 5.2   Examples of power dividers/combiners. (a) Resistive, (b) quadrature hybrid, and (c) Wilkinson.

are matched. Unequal power division is done by having different impedances in the same stage for ports 2 and 3. A power divider can also be made 1 to N where $N > 2$.

5.5   DISTRIBUTED VERSUS CENTRALIZED AMPLIFICATION

Distributed amplification has T/R modules at each antenna element, while centralized amplification has one transmitter and receiver with a corporate feed that distributes the signal to or from each element. Another approach between these extremes is to place amplification at the subarray ports. Centralized amplification has the following attributes [8]:

- The beamformer adds noise to the signals passing through it. Distributed amplification boosts the signal relative to the noise before passing through the noisy receive beamformer, whereas centralized amplification does not.
- Phase shifters and attenuators must handle significantly higher power than a distributed system.
- A centralized amplifier sends the same phase noise to each antenna element. The phase noise arises from power supply interaction with the T/R modules in addition to the inherent amplifier noise. Distributed amplification reduces the phase noise contribution by the number of T/R modules.
- If the amplifier fails, then the whole array fails A failure of a single distributed amplifier has little effect on the performance of an array. A centralized amplifier can be made by power combining several lower power amplifiers to minimize

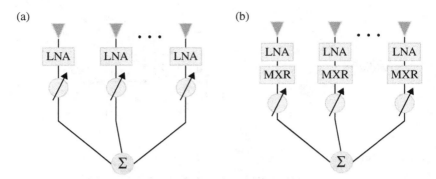

FIGURE 5.3   AESA beamforming (a) RF and (b) IF.

the impact of a single amplifier's failure. Output power and efficiency degradation created by the insertion loss of the output power combiner is a significant drawback for solid-state centralized amplifiers.

- Filters that reduce electromagnetic interference (EMI) from other systems make distributed T/R modules expensive and complex. A centralized system needs only one EMI filter that does not have to fit inside the unit cell. Distributed amplification requires a filter in every T/R module. Distributed filtering will typically have a higher insertion loss due to the packaging size restrictions. The filter's loss adds noise and degrades the output power depending on its placement in the array.

Figure 5.3 shows two approaches to combining received signals in an array [9]. RF beamforming amplifies the signals prior to the phase shifters and combiner in order to compensate for network losses. As a result, the array has good noise performance and impedance matching over a wide frequency band. IF beamforming down-converts the RF signal to an intermediate frequency (IF) prior to the phase shifters and the combiner (Fig. 5.4). Thus, the feed network operates at a lower frequency (IF instead of RF), which simplifies the feed-network design and reduces losses. Low-noise amplifiers (LNAs) can be excluded from the block diagram or, alternately, can be inserted after mixers, thus operating at the IF.

5.6   BLASS MATRIX

A Blass matrix has M beams formed from N elements [10]. A beam port samples the signal from each element as shown in Fig. 5.4. The equally spaced signal couplers create a constant phase shift between elements that steers the beam to a desired angle. All transmission lines terminate in matched loads to prevent reflections. If $\psi_{m,n}$ is the phase length from beam port m to element n, then the phase difference between any two adjacent element is

$$\Delta\psi_m = \psi_{m,n+1} - \psi_{m,n} = \psi_{m,n} - \psi_{m,n-1} \tag{5.15}$$

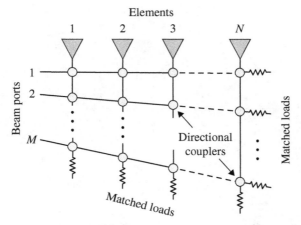

FIGURE 5.4 Diagram of a Blass matrix.

This progressive linear phase shift steers beam m to an angle of

$$u_s = -\frac{\Delta\psi_m}{kd} = -\frac{c\Delta\psi_m}{\omega d} \tag{5.16}$$

The beam steering in (5.16) is a function of frequency, so the Blass matrix is not broadband, because it is vulnerable to beam squint.

5.7 BUTLER MATRIX

The Butler matrix [11] transforms the element signals in a $N = 2^M$ uniform linear array into $N = 2^M$ orthogonal beams Figure 5.5. A basic form of a Butler matrix is the 3 dB quadrature hybrid coupler. The N beams have a linear phase shift given by [12]

$$\delta_{sn} = (2n-1)\pi/N \quad \text{for } n = 1, 2, \ldots, N \tag{5.17}$$

As a result, the beam peaks occur at

$$\delta_{sn} = (2n-1)\pi/N \quad \text{for } n = 1, 2, \ldots, N \tag{5.18}$$

The angular separation between the two outer beams is [12].

$$\text{Angular coverage} = 2\sin^{-1}\left[\frac{c}{2df}\left(1-\frac{1}{N}\right)\right] \tag{5.19}$$

The orthogonal beams overlap at the −3.9 dB level and have the full gain of the array. As with the Blass matrix, the Butler matrix exhibits beam squint. The angular coverage decreases as frequency increases).

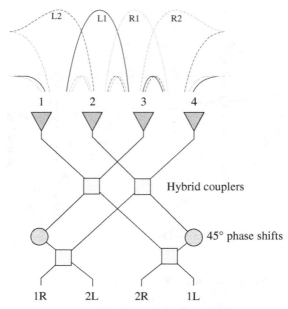

FIGURE 5.5 Diagram of a four port Butler matrix and its four orthogonal beams.

5.8 LENSES

A space-fed lens transmits/receives at beam ports behind the lens. Figure 5.6 is the waveguide lens for the Nike AJAX MPA-4 radar. This lens passively collimates the signals by delaying the shorter propagation paths, so that all paths are equal. The feed sits at the single focal point behind the lens (out of sight at the bottom of the picture).

A bootlace lens replaces the waveguide lens in Figure 5.6 with two back-to-back arrays with elements facing in opposite directions [13]. An element on one side receives the signal and sends it to an element on the other side for transmission. The shape of the input and output arrays, the length of the transmission lines between the input and output elements, and the amplitude and/or phase weights between the receive and transmit elements determine the performance of the lens. A bootlace lens has two conjugate focal points in addition to an on axis focal point. The feed has either one antenna placed at a focus or multiple antennas along a focal arc to get multiple beams. When the lens has a radius of $2R$, then the focal points lie on a circle having radius R [14]. If the transmission lines are the same length, then it is independent of frequency.

A Rotman lens [15] is a bootlace lens derivative with M beam ports (x_{bm}, y_{bm}) placed on an arc that passes through its three focal points (Fig. 5.7). Elements along the curved portion of the lens (x_{pn}, y_{pn}) transmit to or receive from the beam ports. The array elements are at (x_n, y_n). Each beam port has a beam pointing in a predetermined direction. Each beam has the directivity of an N element array. Figure 5.8 is a microstrip version of a Rotman lens made on FR4 [17]. The lens has 13 beam ports that connect to six elements. Its measured patterns at 2.5 GHz are shown in Figure 5.9.

FIGURE 5.6   Waveguide lens antenna for the Nike AJAX MPA-4 radar.

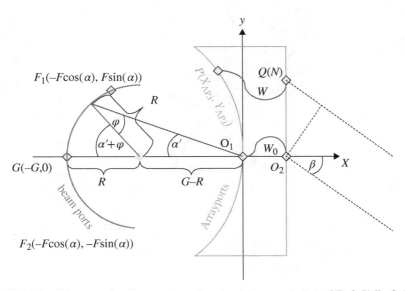

FIGURE 5.7   Diagram of a Rotman lens. Reprinted by permission of Ref. [16]; © 2010 IEEE.

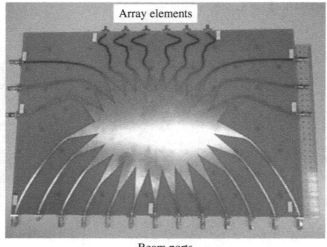

FIGURE 5.8 Microstrip version of a Rotman lens. Reprinted by permission of Ref. [17]; © 2013 IEEE.

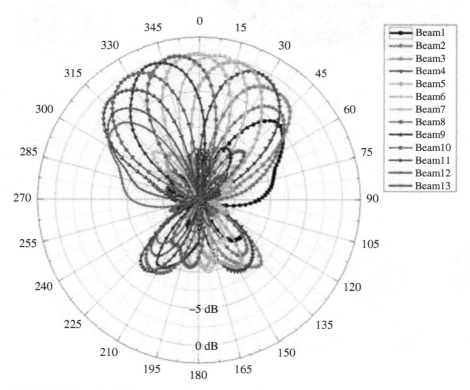

FIGURE 5.9 Measured antenna patterns at 2.5 GHz a Rotman lens array. Reprinted by permission of Ref. [17]; © 2013 IEEE.

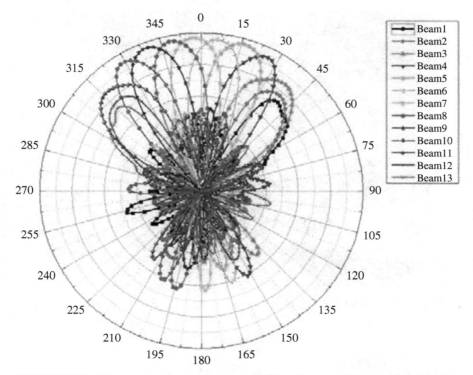

FIGURE 5.10 Measured antenna patterns at 4 GHz a Rotman lens array. Reprinted by permission of Ref. [17]; © 2013 IEEE.

These beams do not squint with frequency as shown by the measured 4 GHz patterns in Figure 5.10.

The AN/MPQ-53 Patriot (Phased Array Tracking Radar to Intercept on Target) radar performs surveillance; identification friend, or foe (IFF); tracking and guidance; and electronic counter measures (ECMs) for the Patriot tactical air defence missile system [18]. The main array (Fig. 5.11) is a circular planar array that is 2.44 m diameter and has 5161-element elements. The other, smaller arrays are for target detection and tracking, missile guidance, sidelobe cancelling, and IFF functions. This space-fed lens antenna can have up to 32 different beams. The search sector is 90°, while the track sector is 120°. The radar can track up to 100 targets and guide up to nine missiles simultaneously.

5.9 REFLECTARRAYS

Reflectarrays are a cross between a reflector antenna and an array antenna [20]. A reflectarray has a feed (often a horn) that transmit/receives to a surface that is a planar array antenna (Fig. 5.12). The elements receive the signal, properly weight the

FIGURE 5.11   AN/MPQ-53 Patriot radar system [19]. Courtesy of Missile Defense Agency.

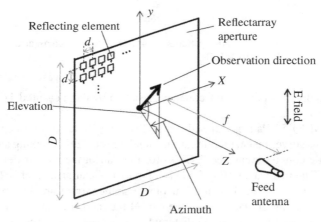

FIGURE 5.12   Diagram of reflect array. Reprinted by permission of Ref. [21]; © 2011 IEEE.

signal, then retransmit it as a collimated beam. Reflectarrays are low-weight and low-profile, are easy to manufacture, have good efficiency, and are high gain [22]. Like a lens antenna, the reflect array eliminates the distribution losses associated with a corporate feed network. A large 60 GHz electronically reconfigurable reflect array with 160×160 elements is shown in Figure 5.13 [21]. Since the reflector was so large, the reflecting elements had to be simple and easy to control. Consequently, the reflecting element was a microstrip patch and a one-bit PIN diode phase shifter.

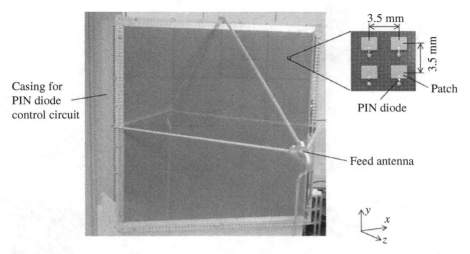

FIGURE 5.13 Experimental reflect array. Reprinted by permission of Ref. [21]; © 2011 IEEE.

5.10 DIGITAL BEAMFORMING

Digital beamforming (DBF) creates receive and/or transmit beams in software rather than forming the beams through RF or IF analog circuits (Fig. 5.14). High-frequency ADCs at the elements transform the RF signals to digital signals. Forming beams in software is easy. Multiple beams, low sidelobes, null placement, and so on, are possible via mathematical algorithms rather than RF hardware [23]. Like T/R modules, the ADC can be either centralized or distributed [24]. DBF is frequently implemented at the subarray level because the ADCs are expensive and require calibration. A full DBF array has the most flexibility and the least amount of feed hardware, because all weighting and combining is done in software. Calibration of the DBF is usually done by coupling a calibration signal into the feed line from the element. Calibration can also be done using near-field or far-field sources [25]. This signal should have the same amplitude and phase at each element after passing through the ADC. Any amplitude and phase errors are corrected in software. DBF is expensive and requires a lot of space behind each element, so it is usually limited to small arrays. DBF systems are moving up in frequency and becoming more common [26].

Sharing the ADC is possible by multiplexing the element signals onto a single channel. The advantage is the hardware cost and calibration requirements become much more reasonable. The disadvantage is that each signal must be multiplexed then demultiplexed, and the ADC bandwidth increases by approximately N. Figure 5.15 is a diagram of a linear array in which the signals are code division multiplexed then converted to digital signals [27]. The digital signals are separated then detected at baseband.

A 61-channel DBF S-band transmit array having a 10 MHz bandwidth with 16 simultaneous beams was built for mobile satellite communication application (Fig. 5.16) [28]. Figure 5.17 is the block diagram of the DBF transmit array. The

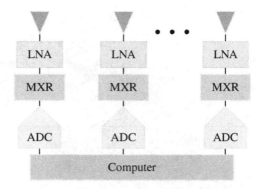

FIGURE 5.14 Diagram of a digital beamforming array.

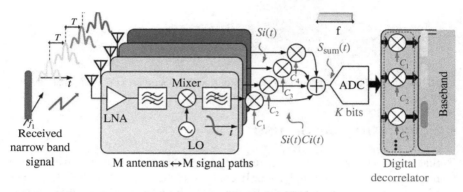

FIGURE 5.15 DBF with on-site coding and one ADC. Reprinted by permission of Ref. [27]; © 2013 IEEE.

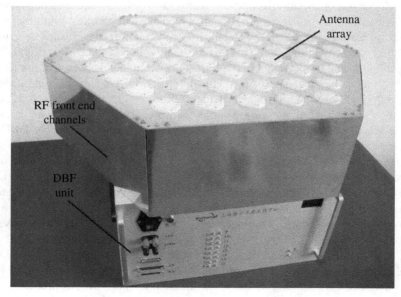

FIGURE 5.16 Photograph of DBF transmitter array. Reprinted by permission of Ref. [28]; © 2009 IEEE.

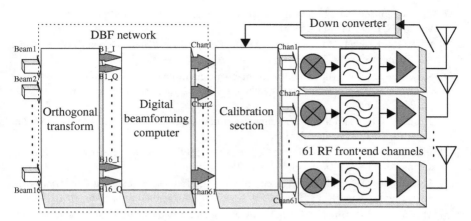

FIGURE 5.17 Block diagram of DBF transmitter antenna. Reprinted by permission of Ref. [28]; © 2009 IEEE.

ADCs have 40 MHz sampling clock rates that convert the RF signals from each beam (IF 70 MHz or 30 MHz) to 12-bit digital signals. A calibration section corrects amplitude and phase errors in the 61 channels. The calibration control module generates a single-tone reference signal that simultaneously propagates through 61 RF frontend channels. After downconversion, the feedback signal is sampled and sent to the error calculation module. Amplitude and phase corrections are calculated from the reference and the feedback signals and combined into complex-valued error correction weights and latched into RAM. Calibration limits errors to less than 0.2 dB and 0.5°. DACs that sample at 40 megasamples/s create 61 calibrated signals that are up-converted to the transmitting frequency (S-band). The RF up-converters have gain variations that are less than 1 dB. Externally generated LO signals fed to RF mixers synchronize the carrier. The 61 RF frontends are mounted on the ground plane to dissipate the power amplifier heat of 5 W/channel.

REFERENCES

[1] Microstrip and stripline design, Analog Devices, MT-094 Tutorial, Rev.0, 01/09, WK, January 2009, pp. 1–7.

[2] D. M. Pozar, *Microwave Engineering* (2nd ed.). New York: John Wiley & Sons, Inc., 1998.

[3] R. Hartley, *RF/microwave PC board design and layout* [online]. Available: http://www. qsl.net/va3iul/. Accessed January 5, 2015.

[4] R. J. P. Douville and D. S. James, "Experimental study of symmetric microstrip bends and their compensation," *IEEE Trans. Microw. Theory Tech.*, vol. 26, no. 3, pp. 175–182, 1978.

[5] J. Stefan. (2007). *Coplanar waveguides (CPW)* [online]. Available: http://qucs. sourceforge.net/tech/node86.html. Accessed December 17, 2014.

[6] Marki Microwave, *Microwave power dividers and couplers tutorial* [online]. Available: http://www.markimicrowave.com/Assets/appnotes/microwave_power_dividers_and_cou plers_primer.pdf. Accessed December 17, 2014.

[7]   E. J. Wilkinson, "An N-way hybrid power divider," *IRE Trans. Microw. Theory Tech.*, vol. 8, no. 1, pp. 116–118, 1960.

[8]   B. A. Kopp, et al., "Transmit/receive modules," *IEEE Trans. Microw. Theory Tech.*, vol. 30, pp. 827–834, 2002.

[9]   S. L. Loyka, "The influence of electromagnetic environment on operation of active array antennas: analysis and simulation techniques," *IEEE Antennas Propagat. Mag.*, vol. 41, pp. 23–39, 1999.

[10] J. Blass, "The multidirectional antenna—a new approach to stacked beams," *IRE Int. Conv. Rec.*, Pt. 1, pp. 48–50, 1960.

[11] J. Butler and R. Lowe, "Beam-forming matrix simplifies design of electronically scanned antennas," *Electron. Des.*, vol. 9, pp. 170–173, 1961.

[12] J. L. Butler, "Digital, matrix, and intermediate-frequency scanning," in *Microwave Scanning Antennas*, R. C. Hansen, Ed. New York: Academic Press, 1966, pp. 217–288.

[13] M. Kales and R. Brown, "Design considerations for two-dimensional symmetric bootlace lenses," *IEEE Trans. Antennas Propag.*, vol. 13, pp. 521–528, 1965.

[14] H. Gent, "The bootlace aerial," *R. Radar Est. J.*, vol. 40, pp. 47–57, 1957.

[15] W. Rotman and R. Turner, "Wide-angle microwave lens for line source applications," *IEEE Trans. Antennas Propag.*, vol. 11, pp. 623–632, 1963.

[16] A. Lambrecht, et al., "True-time-delay beamforming with a Rotman-lens for ultra wideband antenna systems," *IEEE Trans. Antennas Propag.*, vol. 58, pp. 3189–3195, 2010.

[17] P. Chiang, et al., "Implementation of direction-of-arrival estimation using Rotman lens array antenna," International Symposium on Electromagnetic Theory, Hiroshima, Japan, May 2013, pp. 855–859.

[18] *Patriot radar AN/MPQ-53* [online]. Available: http://www.radartutorial.eu/19.kartei/karte423.en.html. Accessed January 5, 2014.

[19] http://www.mda.mil/global/images/system/pac-3/p39.jpg. Accessed December 17, 2014.

[20] D. Berry, et al., "The reflectarray antenna," *IEEE Trans. Antennas Propag.*, vol. 11, pp. 645–651, 1963.

[21] H. Kamoda, et al., "60-GHz electronically reconfigurable large reflectarray using single bit phase shifters," *IEEE Trans. Antennas Propag.*, vol. 59, pp. 2524–2531, 2011.

[22] P. Nayeri, et al., "Beam-scanning reflectarray antennas: a technical overview and state-of-the-art", 2015.

[23] H. Steyskal, "Digital beamforming at Rome laboratory," *Microw. J.*, vol. 39, pp. 100–124, 1996.

[24] J. Frank and J. D. Richards, "Phased array radar antennas," in *Radar Handbook*, M. I. Skolnick, Ed. New York: McGraw Hill, 2008, pp. 13.1–13.74.

[25] L. Pettersson, et al., "An experimental S-band digital beamforming antenna," *IEEE Aerosp. Electron. Syst. Mag.*, vol. 12, pp. 19–29, 1997.

[26] E. Brookner, "Phased arrays and radars—past, present, and future," *Microw. J.*, vol. 49, pp. 1–17, 2006.

[27] E. A. Alwan, et al., "Low cost, power efficient, on-site coding receiver (OSCR) for ultra-wideband digital beamforming," IEEE Phased Array Conference, Waltham, MA, USA, October 2013, pp. 202–206.

[28] L. Guang, et al., "61-Channel digital beamforming (DBF) transmitter array," Asia-Pacific Microwave Conference, Singapore, December 2009, pp. 739–742.

6

ACTIVE ELECTRONICALLY SCANNED ARRAY TECHNOLOGY

An active electronically scanned array (AESA) amplifies and phase shifts the transmit and receive signals by placing active components at the elements. Figure 6.1 is a diagram of a typical T/R (transmit/receive) module. It may operate in half- or full-duplex mode (sequential or simultaneous transmit and receive), depending on the application. The transmit and receive channels must be isolated, and the transmit amplifier must be protected from reflections due to the changes in impedance during beam scanning. Types of isolating devices include circulators and/or switches (Fig. 6.1 has one of each). In order to reduce components, the transmit and receive signals often share the attenuator and phase shifter that weight the amplitude and phase of the signals. The low-noise amplifier (LNA) establishes a low-noise figure while amplifying the receive signal. A limiter caps the signal amplitude in order to protect the sensitive LNA from large input signals. Power amplifiers (PAs) in the transmit channel compensate for the losses in the feed network. However, increasing the power amplifier output power too much has the negative effect of increasing the heat production which could lead to an expensive and complex cooling requirement.

This chapter introduces T/R modules and associated technology, performance terms, and characteristics. The nonlinear effects of power amplifiers are discussed in terms of their impact on array performance. Since switches play an important role in many timed array designs, they are presented in some detail. Finally, the chapter ends with a presentation of phase shifters and attenuators.

Timed Arrays: Wideband and Time Varying Antenna Arrays, First Edition. Randy L. Haupt.
© 2015 John Wiley & Sons, Inc. Published 2015 by John Wiley & Sons, Inc.

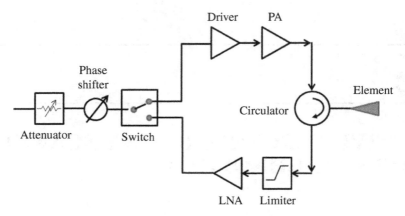

FIGURE 6.1   Diagram of a T/R module.

6.1   SEMICONDUCTOR TECHNOLOGY FOR T/R MODULES

Monolithic microwave integrated circuit (MMIC) T/R modules technology sprang to life in the 1960s. MMICs have passive components (inductors, capacitors, resistors, and transmission lines) fabricated on a semiconductor wafer along with the active devices (transistors, diodes, etc.) [1]. Hybrid assemblies integrate discrete capacitors, inductors, and resistors with the transistors on the MMIC. Si dominates lower frequency applications, while gallium arsenide (GaAs) dominates high frequencies. GaAs is a dielectric with low microwave transmission losses compared to Si. GaAs is more expensive than Si because of its increased processing complexity, smaller wafer diameters, and smaller industry infrastructure. It has less noise, has a relative insensitivity to heat, is highly resistive (good for isolating components), and handles high power. Metal semiconductor field-effect transistors (MESFETs), pseudomorphic high electron-mobility transistors (PHEMTs), and heterojunction bipolar transistors (HBTs) are the three most commonly used GaAs transistors. Other semiconductor technologies, such as indium phosphide (InP), silicon carbide (SiC), and gallium nitride (GaN), have performance advantages over GaAs, but they currently have higher manufacturing costs. Because of the large number of T/R modules required for a typical system, cost rather than performance is the main impediment to fielding phased arrays. Si is usually the substrate of choice, because manufacturing is cheap in large quantities, and it integrates well with other functions. Gallium nitride (GaN) is gradually replacing GaAs as the technology of choice for high-power amplifiers, because GaN has 5–10 times the power performance of GaAs for the same die area [2]. A GaN wafer may cost twice as much as a GaAs wafer, but the GaN MMIC is 1/3 to 1/4 the size of the GaAs MMIC. The resulting GaN solution is only 50–66% the dollars per RF watt generated [3]. GaN material requires a high breakdown voltage, resulting in a power density good for high-power applications. Power density is the power divided by the transistor size. The high voltage breakdown associated with GaN also contributes to increased robustness and survivability under

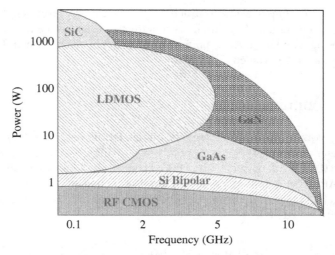

FIGURE 6.2   Semiconductor technology for T/R modules [4].

a harsh RF environment. Figure 6.2 shows the frequency range and power-handling capabilities of several different semiconductor technologies [4]. AESAs at L-band, S-band, or X-band use SiC, GaAs, GaN, and indium phosphide (InP). Silicon germanium (SiGe) is a low-power alternative (<1 W at X-band). Si lateral double-diffused metal oxide semiconductor (LDMOS) is a technology that works well at UHF but is not suitable at higher frequencies. Table 6.1 compares the leading high-power amplifier technologies based on a high-power amplifier delivering 20 W of RF output power.

Traditional MMIC power amplifiers in T/R modules are class AB amplifiers with maximum power-added efficiency (PAE) between 50 and 78.5%, depending on operating frequency—higher frequency has lower efficiency [3]. In practice, the PAE ranges from 35 to 45% for GaAs-based amplifiers and from 50 to 60% for large periphery GaN-based amplifiers. GaN has a high breakdown voltage that allows the transistor amplifier to operate at high power and high efficiency compared to GaAs at the same power levels. This is largely due to the fact that GaAs requires lossy

TABLE 6.1   Semiconductor Technologies for T/R Modules[a]

Semiconductor technology	DC bias voltage (V)	Power density (W/mm)	Peak current (A)	Chip area (mm)^2	Thermal conductivity (W/mK)	Cost ($/W)
GaN HEMT	24–48	>5	2.5	17	390	12–14
HV GaAs	15–24	2	5	25	44	15–20
LV GaAs	8–10	0.7	6	40	44	18–20

[a]Ref. [3].

power combining of more elements to achieve the same power levels. High-efficiency amplifiers run cooler than low-efficiency amplifiers. Heat must be removed through either air or liquid cooling. Air is preferred because it is much cheaper. Liquids and electronics should be avoided if possible.

6.2 T/R MODULE LAYOUT

The first monolithic microwave integrated circuit T/R module (Fig. 6.3) grew out of the molecular electronics for radar applications (MERAs) program in 1964 [6]. It transmitted 0.6 W peak power at 9.0 GHz and had a noise figure of 12 dB with a gain of 14 dB. Amplifier gain is the ratio of the power output to the power input in the linear operating range of the amplifier. The noise figure (NF) is defined as the noise factor (F) expressed in dB.

$$\text{NF} = 10\log_{10}(F) = 10\log_{10}\left(\frac{\text{SNR}_{\text{Input}}}{\text{SNR}_{\text{Output}}}\right) \tag{6.1}$$

The $7.1 \times 2.5 \times 0.8$ cm module weighed 0.95 ounces. It was fabricated on high-density alumina and high resistivity Si, using thin-film techniques. Ceramic substrates made from thermally conductive beryllium oxide or aluminum oxide form internal mounting media for the MMIC devices. Device and circuit modeling significantly improved in the 1980s, and resulted in designs that made use of microelectronics automated assembly and test techniques. GaAs MMIC devices in conjunction with other Si devices combine the array controller, power system, cooling system, and RF distribution network into one package. In 1987, a new generation of T/R modules sprang from the advanced tactical fighter (ATF) program [7]. These modules used GaAs MMIC technology in order to reduce the size and cost (Fig. 6.4). The AN/APG-77 radar antenna used 2000 of these T/R modules in an array that scanned ±60°. AESAs

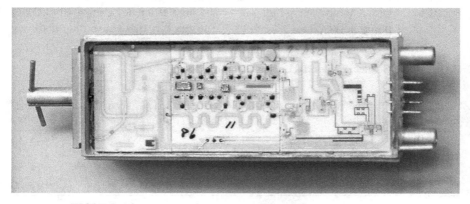

FIGURE 6.3 Picture of the MERA T/R module [5] © 2010 Wiley.

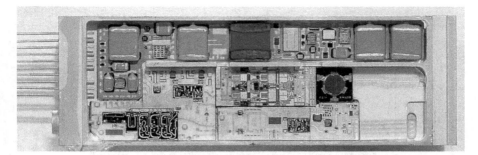

FIGURE 6.4 Picture of the ATF T/R module [5] © 2010 Wiley.

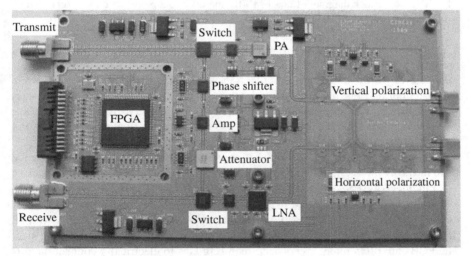

FIGURE 6.5 Photograph of the T/R module for the X-band phase-tilt radar array. Reprinted by permission of Ref. [8]; © 2012 IEEE.

predominantly used GaAs pseudomorphic high electron mobility transistor technology in T/R modules in the 1990s and early 2000s [2].

A photograph of the T/R module developed for the X-band Phase-Tilt radar array is shown in Figure 6.5 [8]. The module was designed using commercial off-the-shelf GaAs MMICs. A custom PIN diode switch connects to either vertical or horizontal polarized elements. It operates at 9.36 GHz and has a 1.0 GHz bandwidth. The transmit and receive channels share a 6-bit digital attenuator and 6-bit digital phase shifter. An FPGA issues commands and RF switching signals. The transmit channel has 30 dB of gain at room temperature with a maximum peak output power of 1.25 W when the module operates in compression. A 30% duty cycle protects the power amplifier from excessive heat. The receive channel has a net gain of 29 dB and noise figure of 4.3 dB.

Figure 6.6 shows the layout of a two-stage 3.1–3.5 GHz high-voltage GaN-on-SiC high electron mobility transistor (HEMT) power amplifier [9]. GaN amplifiers

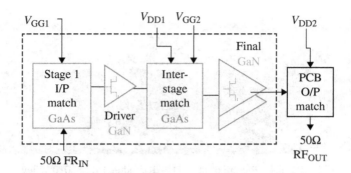

FIGURE 6.6 Block diagram of quasi-MMIC approach to a power amplifier. Reprinted by permission of Ref. [9]; © 2013 IEEE.

decrease the package size due to their high power density. Two amplifying stages produce the desired gain. The area of the driver and final stage GaN-on-SiC chips is minimized by only including the active transistors and associated bond pads, while the matching and biasing circuitry is integrated on lower-cost GaAs. The low-loss GaAs substrate and gold metallization produce high-density, high-quality passive circuitry at much lower cost than a SiC substrate. Since the final stage output is not 50Ω, external matching is placed on the PCB.

The modules in Figures 6.3 and 6.4 are called "bricks," because they protrude from the array plane and look like stacked bricks [10]. Each element has one brick or T/R module. An AESA consists of a stack of these bricks. Today, the tile approach to assembling AESAs is preferred over the brick approach, because it has much lower manufacturing costs. An AESA tile is a subarray or subsection of an AESA with all the T/R MMICS laid out on the plane of the antenna array on the nonradiating side. The PCBs are often many layers deep requiring vias to electrically connect the different layers. Placing all the components on one side of the PCB tremendously simplifies the manufacturing process. Figure 6.7 is an example of tile architecture [11]. The front PCB (antenna aperture) has the elements, while the PCBs below contain the components and feed network. Other layers are sandwiched in between. A typical large PCB is 18×24 in., so a large array has many of these PCB panels placed side by side.

6.3 AMPLIFIERS

Amplifiers increase the signal power. Ideally, this increase should be linear, but in reality, the output power is a nonlinear function of the input power as shown by the P_{out} vs. P_{in} curve in Figure 6.8. A general memory-less model for an amplifier with input, v_{in}, and output, v_{out}, at the fundamental frequency, f_c, is given by [12]

$$v_{out,n}(t) = g\left[v_{in,n}(t)\right]\cos\left\{2\pi f_c t + \xi\left[v_{in,n}(t)\right]\right\} \qquad (6.2)$$

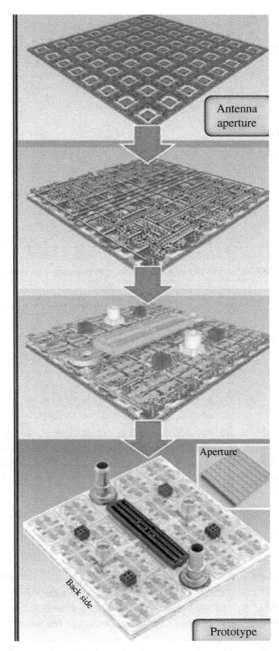

FIGURE 6.7 Tile architecture with array elements on the front and components on the back. Reprinted by permission of Ref. [11]; © 2013 IEEE.

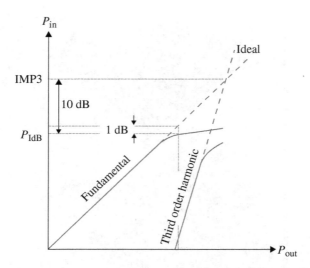

FIGURE 6.8   Output power versus input power for an amplifier.

where g and ξ are the AM/AM and AM/PM distortion functions (AM is amplitude modulation and PM is phase modulation). Memory-less models ignore thermal effects of the active devices, so the output is only a function of the current input and not past inputs. The nonlinear polynomial model of a memory-less amplifier is given by

$$v_{out}(t) = k_0 + k_1 v_{in}(t) + k_2 v_{in}^2(t) + k_3 v_{in}^3(t) + \cdots \tag{6.3}$$

Usually, a third-order model is sufficient and all terms above k_3 are ignored. Although this polynomial model is simple and easy to manipulate in mathematical expressions, for more realistic systems, the Rapp model for solid-state amplifiers is used [13]

$$v_{out}(t) = \frac{v_{in}(t)}{\left[1 + \left(|v_{in}(t)|/v_{sat}\right)^{2\varsigma}\right]^{1/(2\varsigma)}} \tag{6.4}$$

where v_{sat} is the saturation voltage of the amplifier and ς is the smoothness factor (determined from measurements) that sets the smoothness of the transition from linear to saturation states. Small values of ς produce a smoother transition than large values. The model assumes linear performance until near the saturation point when it transitions to an almost constant saturated output.

Spectral regrowth results when a power amplifier operates at high power (non-linear region) and generates unwanted spectral sidebands. These unwanted sidebands appear as interference in nearby channels and are difficult to filter. Adjacent channel interference is characterized by the adjacent channel power ratio (ACPR)

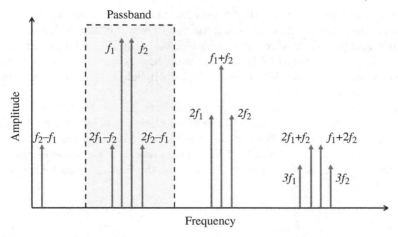

FIGURE 6.9   Intermodulation products resulting from two tones input to the amplifier.

which is the power in the main channel divided by the power in the lower plus upper adjacent channels [14].

Spurious signals arise in the amplifier output when k_n, $n \geq 2$ and v_{in} become large enough relative to $k_0 + k_1 v_{in}$. For example, harmonic distortion occurs when a single-frequency input signal at f_c produces spurious output signals at the harmonic frequencies $2f_c, 3f_c, \ldots$ (Fig. 6.9). Harmonics are usually measured in dBc where "c" indicates carrier. Another example are intermodulation products (IMPs) which are the frequency components (other than the harmonics) that result when two closely spaced equal amplitude tones at f_1 and f_2 are input to an amplifier (Fig. 6.9). IMPs result at the frequencies:

$$f_{IMP} = \pm mf_1 \pm nf_2, \quad m, n \text{ are integers} \tag{6.5}$$

when the tone amplitudes are of sufficient magnitude [15].

As shown in Figure 6.9, the third-order intermodulation products (IMP3s) occur when $m + n = 3$ or $2f_1 - f_2$ and $2f_2 - f_1$. They are the closest frequency components to the input tones, making them difficult to filter. They often appear in the passband of the signal. Additional IMPs and harmonics may also enter the passband of a wideband system. Higher-order IMPs increase in power much faster than the fundamental as shown by the third-order harmonic in Figure 6.8. The second harmonic and the second-order intermodulation products' power increases two times (in dBm) faster than the fundamental, while the third-order harmonics increase three times as fast. The *third-order intercept* occurs when the ideal linear third-order IMP equals the ideal linear, uncompressed fundamental output. IMP3 approximately equals $P_{1dB} + 10dB$ [16].

The compression point is another descriptive measure of the amplifier linearity. A device operates in a linear region when a 1 dB increase in input power causes a 1 dB increase in the output power. At some input power, however, a 1 dB increase in input power results in less than a 1 dB increase in the output power. The 1 dB compression point (P_{1dB}) occurs when the output power of the device is 1 dB less than it would be if the device were linear (see Fig. 6.8). P_{1dB} is important, because it is a figure of merit (FOM) for the maximum power for linear operation. Typically, the maximum linear power is 1–6 dB below P_{1dB}, depending on the signal modulation.

An LNA amplifies the received signal while having a low NF [17]. LNAs immediately follow the antenna element in order to minimize noise. A band pass filter between the element and LNA prevents interference from adjacent bands but adds noise to the system. The gain of the LNA reduces receive noise power in subsequent amplifier stages of the receive chain as shown by the Friis equation for a system of cascaded amplifiers:

$$F_{system} = F_1 + \frac{F_2 - 1}{G_1} + \cdots + \frac{F_N - 1}{G_1 G_2 \cdots G_{N-1}} \tag{6.6}$$

where F_n is the noise factor and G_n the gain of amplifier n, while $n = 1$ the LNA.

Normally, an array has a single beam, and the amplifiers are designed to be linear over the power levels and operating frequencies. A multibeam array transmitting M signals through M beams in M different directions at M frequencies may push amplifiers to operate in the nonlinear region for lower power levels than if a single beam were used. Each signal/beam has its own feed network to the elements as shown in Figure 6.10. Phase shifters in a feed network steer the beams associated with that feed network. The signals from all the feed networks are first summed before amplifying then radiating from the element. As a result, the array factor is written as follows:

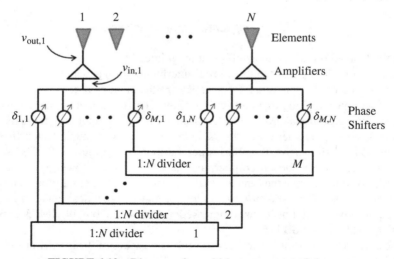

FIGURE 6.10   Diagram of a multi-beam corporate fed array.

$$AF = \sum_{n=1}^{N}\sum_{m=1}^{M} v_{\text{out},n} e^{j\frac{2\pi f_m}{c}(x_n u + y_n v)} \tag{6.7}$$

where in Figure 6.10

$$\delta_{mn} = n\frac{2\pi d f_m}{c}\sin\theta_m \tag{6.8}$$

If a single tone (f_c) enters the T/R module at element n of a linear array

$$v_{\text{in},n} = a_n \cos(\omega t - \delta_n) \tag{6.9}$$

then the output from the T/R module at element n when modeled by a third-order polynomial is

$$v_{\text{out},n} = 0.5k_2 a_n^2 + (k_1 a_n + 0.75 k_3 a_n^3)\cos(\omega t - \delta_n) + 0.5k_2 a_n^2 \cos[2(\omega t - \delta_n)]$$
$$+ 0.25 k_3 a_n^3 \cos[3(\omega t - \delta_n)] \tag{6.10}$$

Substituting (6.10) into (6.7) with one beam at one frequency results in

$$AF = \underbrace{\sum_{m=1}^{N}\left(k_1 a_n + 0.75 k_3 a_n^3\right) e^{j(\omega_c t + k x_n u + \delta_n)}}_{\text{fundamental}} + \underbrace{\sum_{m=1}^{N} k_2 a_n^2 e^{j2(\omega_c t + k x_n u + \delta_n)}}_{\text{second harmonic}} + \underbrace{\sum_{m=1}^{N} 0.25 k_3 a_n^3 e^{j3\,0\omega_c t + k x_n u + \delta_n}}_{\text{third harmonic}}$$

$$\tag{6.11}$$

The second and third harmonic beam amplitude tapers are squared and cubed versions of the original fundamental amplitude distribution and all beams point in the same direction, because the phase for each beam at mf_m is $m\delta_n$. Realistic third-order polynomial models have a negative value for k_3 which implies the fundamental beam gain is reduced at high power levels [18].

The IMPs at the amplifiers are coherent and form beams in directions that are determined by the phase of the signals. Some of the IMP beams appear at angles far from the desired main beams. If the main beam has a low sidelobe taper, then the IMP beams have even lower sidelobe tapers. The IMP beams have a quadratic phase error resulting from the AM–PM conversion. Variations in the nonlinear amplifier characteristics induce errors in the main beams and the IMP beams. The magnitude of these errors is larger for the IMP beams, since the difference between higher-order terms tends to be greater than the difference between lower-order terms in the models of two slightly different nonlinear amplifiers

Now, consider a two-tone input with each tone transmitted in different directions. The signal input to the T/R module at element n is

$$v_{\text{in},n} = a_n [\cos(\omega_1 t + \delta_{1n}) + \cos(\omega_2 t + \delta_{2n})] \tag{6.12}$$

The power amplifier output at element n is found by substituting (6.12) into (6.3) to obtain the following [18]:

$$
\begin{aligned}
v_{\text{out},n} = k_2 a_n^2 && \text{dc} && (6.13)\\
+ k_2 a_n^2 \cos[(\omega_1 - \omega_2)t + (\delta_{1n} - \delta_{2n})] && \text{difference}\\
+ (k_1 a_n + 2.25 k_3 a_n^2)\cos(\omega_1 t + \delta_{1n}) && f_1 \text{ fundamental}\\
+ (k_1 a_n + 2.25 k_3 a_n^2)\cos(\omega_2 t + \delta_{2n}) && f_2 \text{ fundamental}\\
+ 0.75 k_3 a_n^3 \cos[(2\omega_1 - \omega_2)t + (2\delta_{1n} - \delta_{2n})] && 2f_1 - f_2 \text{ IMP}\\
+ 0.75 k_3 a_n^3 \cos[(2\omega_2 - \omega_1)t + (2\delta_{2n} - \delta_{1n})] && 2f_2 - f_1 \text{ IMP}\\
+ k_2 a_n^2 \cos[(\omega_1 + \omega_2)t + (\delta_{1n} + \delta_{2n})] && f_1 + f_2 \text{ IMP}\\
+ 0.5 k_2 a_n^2 \cos(2\omega_1 t + 2\delta_{1n}) && f_1 \text{ second harmonic}\\
+ 0.5 k_2 a_n^2 \cos(2\omega_2 t + 2\delta_{2n}) && f_2 \text{ second harmonic}\\
+ 0.75 k_3 a_n^3 \cos[(2\omega_1 + \omega_2)t + (2\delta_{1n} + \delta_{2n})] && 2f_1 + f_2 \text{ IMP}\\
+ 0.75 k_3 a_n^3 \cos[(2\omega_2 + \omega_1)t + (2\delta_{2n} + \delta_{1n})] && 2f_2 + f_1 \text{ IMP}\\
+ 0.25 k_3 a_n^3 \cos(3\omega_1 t + 3\delta_{1n}) && f_1 \text{ third harmonic}\\
+ 0.25 k_3 a_n^3 \cos(3\omega_2 t + 3\delta_{2n}) && f_2 \text{ third harmonic}
\end{aligned}
$$

The beam pointing direction for each of these terms is shown in Table 6.2 under the u column ($u_n = \sin\theta_n$ for a linear array). The pointing directions of the IMP beams depend on the frequency of the tones and the beam pointing directions.

Consider a dual-beam uniform linear array with a 6 GHz carrier scanned to 30°, an 8.4 GHz carrier scanned to 60° and third-order polynomial amplifier model. Both beams share the same high power amplifier. Figure 6.11 is a plot of the T/R module relative output spectrum, while Figure 6.12 shows the beam magnitudes and locations at the various output frequencies. Arrays do not have infinite bandwidth, though. If the array operates between 5.9 and 14.5 GHz, then the number of radiated signals significantly decreases as shown by the plots in Figures 6.13 and 6.14. The system bandwidth is 84% and includes the two fundamentals, third-order difference, second harmonic, and sum signals. All other harmonics and IMPs are rejected, because they are outside the bandwidth. Some of the frequencies do not have corresponding beams, because the beams lie outside of visible space (±90°).

For planar arrays, the equations for the spurious frequencies generated by the T/R module for a two-tone input are the same as for a linear array with the beam steering phase appropriately calculated for azimuth-elevation scanning. Table 6.2 lists the beam pointing directions in u–v space for the corresponding radiated frequencies. As an example, consider a two-beam planar array with a 6 GHz beam steered to $(u_1, v_1) = (0.1, 0.1)$ and an 8.4 GHz beam steered to $(u_2, v_2) = (-0.1, -0.1)$. Figure 6.15 shows the location of the main beams and IMP beams in u–v space calculated using Table 6.2.

Hemmi provides some rules of thumb for calculating the power of the signal emanating from a power amplifier when two tones are input [18]. These results are summarized in Table 6.3 when a third-order polynomial model is used.

TABLE 6.2　Beam Locations When Two Tones Are Transmitted[a]

Frequency	Signal	Beam pointing direction	
		u	v
f_1	f_1 fundamental	u_1	v_1
$2f_1, 3f_1$	f_1 harmonics	u_1	v_1
f_2	f_2 fundamental	u_2	v_2
$2f_2, 3f_2$	f_2 harmonics	u_2	v_2
$2f_1 - f_2$	Third order difference IMP	$\dfrac{2f_1 u_1 - f_2 u_2}{2f_1 - f_2}$	$\dfrac{2f_1 v_1 - f_2 v_2}{2f_1 - f_2}$
$2f_2 - f_1$	Third order difference IMP	$\dfrac{2f_2 u_2 - f_1 u_1}{2f_2 - f_1}$	$\dfrac{2f_2 v_2 - f_1 v_1}{2f_2 - f_1}$
$f_1 - f_2$	Difference	$\dfrac{f_1 u_1 - f_2 u_2}{f_1 - f_2}$	$\dfrac{f_1 v_1 - f_2 v_2}{f_1 - f_2}$
$f_1 + f_2$	Sum	$\dfrac{f_1 u_1 + f_2 u_2}{f_1 + f_2}$	$\dfrac{f_1 v_1 + f_2 v_2}{f_1 + f_2}$
$2f_1 + f_2$	Third order sum IMP	$\dfrac{2f_1 u_1 + f_2 u_2}{2f_1 + f_2}$	$\dfrac{2f_1 v_1 + f_2 v_2}{2f_1 + f_2}$
$2f_2 + f_1$	Third order sum IMP	$\dfrac{2f_2 u_2 + f_1 u_1}{2f_2 + f_1}$	$\dfrac{2f_2 v_2 + f_1 v_1}{2f_2 + f_1}$

[a] Ref. [18].

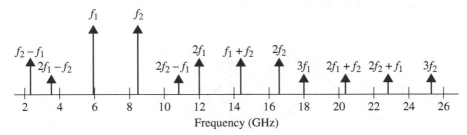

FIGURE 6.11　Amplitude and frequencies at the output of the T/R module when the input has 6 and 8.4 GHz tones.

Sandrin experimentally demonstrated IMP beam generation using a four-element array with a 4×4 Butler matrix feed having traveling wave tubes (TWTs) at the elements (Fig. 6.16) [19]. Two tones (4000 and 4010 MHz) input at ports 2 and 3 of the Butler matrix generated third-order IMPs at 3990 and 4020 MHz. The tone amplitudes were sufficient to drive the TWTs into saturation. Figure 6.17 shows plots of the measured patterns of the fundamentals and the IMPs. In this case, the IMP beams are 14 dB below the desired beams and steered away ±19° from boresight.

The behavior of the IMP beams, as well as the fundamental beams, depends on whether beam steering is done with phase shifters or time delay units [20]. Figure 6.18

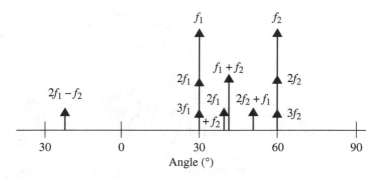

FIGURE 6.12   Magnitude and locations of the beams corresponding to the frequencies in Figure 6.11.

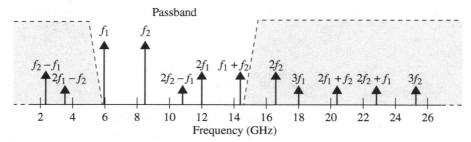

FIGURE 6.13   Pass band spectrum of Figure 6.11.

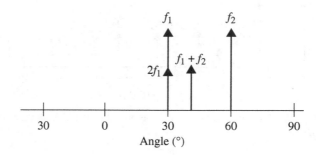

FIGURE 6.14   Magnitude and locations of the beams corresponding to the frequencies in Figure 6.13.

shows elevation cuts of the Relative far-field patterns of a 16×16 array with a square grid spacing of 13.5 cm at the two-tone frequencies of 0.95 ad 1.05 GHz and their third- and fifth-order IMPs when using phase shifters to steer the beams to $\theta_s = -30°$ and $20°$ at the center frequency of 1 GHz. Note the beam squint of all the radiated beams, because the beam steering phase shifts were calculated at 1 GHz. By contrast, Figure 6.19 shows the patterns for the same array when time delay units at the elements

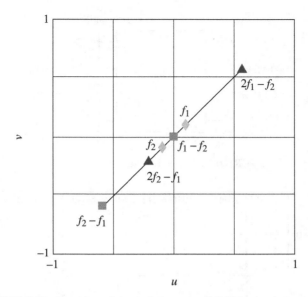

FIGURE 6.15 Location of the main beams and IMP beams in u–v space.

TABLE 6.3 Estimates of Important Power Points[a]

Variable	Definition
$P_{1dB} \propto k_1/k_3$	1 dB compression point
$P_{3OI} \propto k_1^3/k_3$	Third order intercept
$P_{2OI} \propto k_1^2/k_2$	Second order intercept
$P_{f1} = G + P_{in}$ for $P_{f1} < P_{1\,dB}$ (dB)	Fundamental output
$P_{2f_1-f_2} = P_{2f_2-f_1} = 3P_{f1} - 2P_{3OI}$ (dB)	Third order IMP output
$P_{2f_1} = 2P_{f1} - P_{2OI}$ (dB)	Second harmonic output

[a]Ref. [18].

steer the beams instead of phase shifters. In this case, there is no beam squint on any of the beams. The patterns differ from those in Figure 6.18, because the two beams at the same frequency coherently add together resulting in different sidelobe levels.

6.4 SWITCHES

Switches play important roles in timed and phased arrays. An RF switch opens or closes a current path via a mechanical connection or by biasing a semiconductor. The "on" time for a switch is defined by the time from when the control pulse reaches

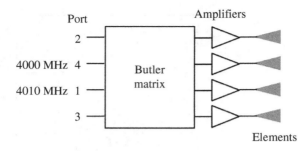

FIGURE 6.16   Experimental multi-beam array.

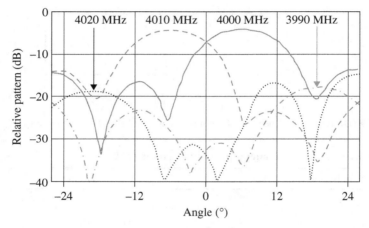

FIGURE 6.17   Pattern measurements of experimental array in Figure 6.16. Mainbeams are at 4000 and 4010 MHz. Third-order IMP beams are at 3990 and 4020 MHz [19].

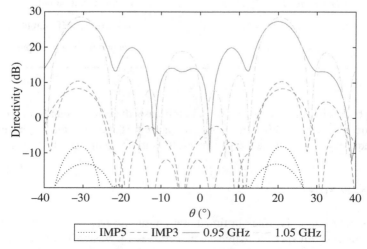

FIGURE 6.18   Relative far field patterns of the array at the two tone frequencies and their third and fifth order IMPs when phase steering [20] © 2013.

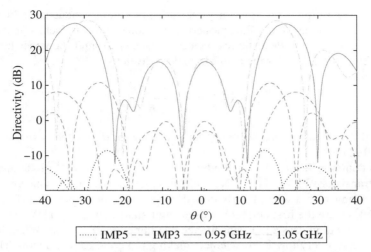

FIGURE 6.19   Relative far field patterns ($\phi = 0°$ cut) of the array at the two tone frequencies and their third and fifth order IMPs when $P_{in} = 5$ dBm time delay steering [20] © 2013.

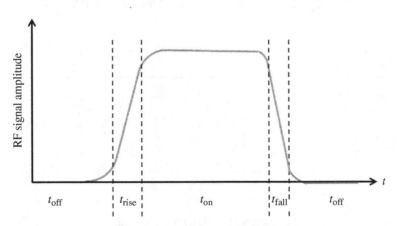

FIGURE 6.20   Turning an RF switch on and off.

50% of its level to the time that the RF signal is at 90% of its peak (Fig. 6.20) [21]. The "off" time for a switch is defined by the time from when the control pulse reaches 50% of its level to the time that the RF signal is at 10% of its peak. Switching time is the larger of the "on" and "off" switch times. Switching speeds of semiconductor switches are on the order of nanoseconds, while switching speeds of mechanical switches are orders of magnitude slower. System constraints limit the applicable type of switch used based on required switching times and power handling. The expected lifetime (measured in number of switch actuations until failure) and power handling (measured in watts) are also important switch attributes.

A low pass switch acts like a resistor (R_{on}) when on and a capacitor (C_{off}) when off, whereas a band pass switch behaves like one capacitor when on and a different capacitor when off. As the frequency increases, the ground inductance, bond-wire inductance, and transmission-line properties become significant factors and complicate the circuit model. Parasitic resistances in the switch limit the upper frequency bound. The switch FOM is the cutoff frequency given by [22]

$$\text{FOM} = \frac{1}{2\pi C_{\text{off}} R_{\text{on}}} \tag{6.14}$$

A rule of thumb for the highest operating frequency for the switch is approximately FOM/10 [23].

Two common forms of semiconductor switches are the PIN diode and field-effect transistor (FET). The PIN diode has heavily doped p-type and n-type regions along with a wide, lightly doped intrinsic region in between (Fig. 6.21). PIN diodes were the first commonly used solid-state switches [21]. When reverse biased, a PIN diode has a low capacitance that impedes RF signals. When forward biased, it has a very low RF resistance, making it a good RF conductor. The FET switch turns on when a voltage applied at the gate increases the conducting channel size beneath the gate and allows current to flow between the source and drain (Fig. 6.22) [21]. Two common versions of the FET switch are MESFET and PHEMT [24].

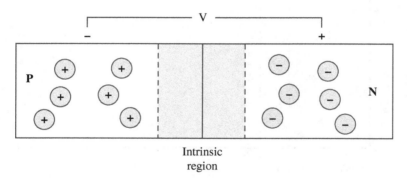

Intrinsic
region

FIGURE 6.21   Diagram of a PIN diode.

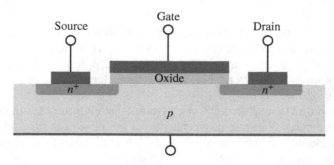

FIGURE 6.22   FET switch.

FETs are the most commonly used semiconductor switch, but the PIN diode still has important applications [21, 24]. FETs are low pass with excellent isolation. Similarly, shunt PIN diodes provide good isolation at high frequencies. Unlike the FET, the PIN diode is current controlled and handles one or more amps of RF current. It consumes a lot of power. Conversely, the FET has very little DC power consumption. PIN diodes are less susceptible to electrostatic discharge and have a higher FOM than an FET. GaAs FET switches use the same planar processes as GaAs amplifiers, so the amplifiers and control functions can be functionally integrated, while GaAs PIN diodes must be vertically processed and require RF chokes to isolate the switching DC power supply from the RF circuitry. An FET has higher "on" resistance and insertion loss than a PIN diode, where insertion loss is the power loss that results from inserting a device in a transmission line or waveguide.

$$IL = 10 \log_{10} \frac{P_{out}}{P_{in}} \tag{6.15}$$

Hybrid PIN diode/FET switches combine the advantages of the two technologies to extend the bandwidth and improve the isolation.

Microelectromechanical systems (MEMS) switches are tiny mechanical switches made on a substrate (Si, quartz, glass) [25]. Figure 6.23 shows two types of MEMS switches in their on and off positions. The cantilever beam in Figure 6.23a and b is anchored to a post on the left, while the other end of the beam suspends above the drain. An applied voltage creates an electrostatic force that pulls the beam down in order to create an electrical path to the drain. The cantilever device travels 6 µm between the on and off states. Figure 6.23c–d show a MEMS membrane switch that has a flexible thin metal membrane anchored to posts at both ends. A potential applied to the bias electrode pulls the membrane down to close the circuit. An ohmic contact is a metal-to-metal connection, while a capacitive contact has a dielectric between the

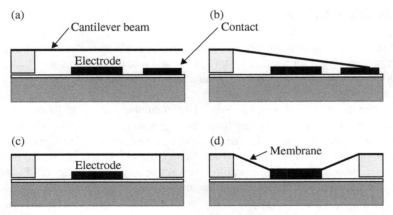

FIGURE 6.23 MEMS switches in the on and off positions. (a) Cantilever off, (b) Cantilever on, (c) Ohmic contact off, and (d) Ohmic contact on.

TABLE 6.4   Characteristics of Switch Technologies[a]

Characteristic	PIN diode	FET	Hybrid	MEMS
Isolation at low frequencies ~100 MHz	Average	Excellent	Excellent	Excellent
Isolation at high frequencies ~18 GHz	Excellent	Average	Excellent	Excellent
Switching time	ns – μs	ns – μs	ns – μs	μs
Insertion loss	Excellent	Good	Average	Low
Repeatability	Excellent	Excellent	Excellent	Good
Power consumption	High	Low	Low	Low
Operating life	High	High	High	Medium
Power handling (W)	≤50	≤50	≤50	≤50
Low frequency limit	100 MHz	DC	300 kHz	DC
High frequency limit	>50 GHz	20 GHz	20 GHz	None

[a]Refs. [24, 27].

two metal contacts. Ohmic switches have a higher bandwidth than capacitive switches. Capacitive MEMS switches handle higher peak power levels and tolerate environmental effects better than ohmic contacts. The FOM or cutoff frequency for MEMS switches is about 9000 GHz [22]. Switching speeds for electrostatically driven capacitor structures are 10 μs. Recently, MEMS switches with piezoelectric films have switching times of 1–2 μs [26]. MEMS switches have low power consumption, low insertion loss, and high isolation like mechanical switches as well as being small, light weight, and low cost like semiconductor switches [24]. Reliability used to be a problem but has dramatically improved in the past few years.

Table 6.4 compares RF switch technologies [21]. Selecting the best switch for an application depends on the characteristics that are most important for meeting system design specifications.

6.5   PHASE SHIFTER

Phase shifter puts the term "phased" in "phased array." A phase shifter changes the phase of a signal passing through it. Its output phase and amplitude are constant across the bandwidth. If a phase shifter has a high insertion loss, then it needs an amplifier. Most phase shifters are reciprocal—they have identical characteristics on transmit and receive. This section describes and compares several different phase shifter technologies.

Fixed line length phase shifters force the signal to take a longer path in order to cause delay, hence a phase shift. The phase of any transmission line equals its length times its propagation constant. A switch (e.g., MEMS) diverts the signal from the shorter reference arm to a longer path corresponding to a phase shifter bit. Phase shift equals the difference in the electrical lengths of the reference and delay arms. The phase shifter digital control signal in Fig. 6.24 is "100" which corresponds to 180°. Since the path lengths determine time delay, the phase shift changes as a function of frequency, so these phase shifters are narrow band.

An in-phase and quadrature (IQ) vector modulator serves as an attenuator as well as a phase shifter [28]. The input signal splits into an in phase (0°) or I channel and quadrature phase (90°) or Q channel as shown in Figure 6.25. Both channels have 1-bit phase shifters that are independently controlled. If the output signal is attenuated by α dB and phase shifted by δ degrees, then the attenuator settings are given by

$$\alpha_I = 20 \log_{10} (10^{-\alpha/20} \cos\delta)$$
$$\alpha_Q = 20 \log_{10} (10^{-\alpha/20} \sin\delta)$$

(6.16)

The 1-bit phase shifter in the two channels is set as indicated in Table 6.5 in order to get the desired phase shift. Generally, an IQ vector modulator performs well over a 3:1 bandwidth.

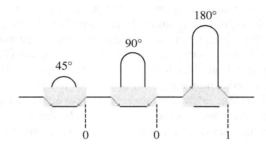

FIGURE 6.24   Three bit switched line phase shifter.

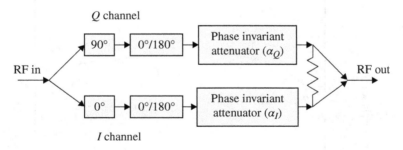

FIGURE 6.25   I-Q vector modulator.

TABLE 6.5   Settings for 1 Bit Phase Shifter in I-Q Vector Modulator

1 bit phase shifter		
I (°)	Q (°)	Desired phase shift (°)
0	0	0–90
180	0	90–180
180	180	180–270
0	180	270–360

A high-pass/low-pass phase shifter is an alternative technology that has nearly constant phase shift over an octave or more [29]. At the input, half the signal goes to a high-pass filter, while the other half goes to a low-pass filter. The cutoff frequencies of the two filter networks lies outside of the phase shift bandwidth. This type of phase shifter takes up less space, because lumped elements are typically used instead of more space-consuming delay lines. Figure 6.26 is a plot of the insertion phase versus frequency for a fictitious phase shifter. In this example, the combined phase is relatively constant over the 4–8 GHz bandwidth (inside box). Figure 6.27 is a photograph of a GaN HEMT high-pass/low-pass phase shifter [29]. This 0°/45° high-pass/low-pass phase shifter has low insertion loss (2.5 dB), good return loss, and an amplitude variation lower than 0.5 dB for the two-phase states over the entire 6–13 GHz operational bandwidth.

Ferrite phase shifters are either reciprocal or nonreciprocal, depending on whether the variable differential phase shift through the device is a function of the direction of propagation [30]. Also, ferrite phase shifters are either latching or nonlatching, depending on whether a continuous holding current sustains the magnetic bias field. Three types of ferrite phase shifters are generally encountered: (1) twin toroid is a latching, nonreciprocal device, (2) dual-mode is a latching, reciprocal device, and (3) the rotary-field is a nonlatching reciprocal device. All use ferrite materials having a square hysteresis loop. The latching phase shifters have either a closed magnetic circuit or a magnetic circuit with very small air gaps. The ferrite permeability is controlled by adjusting the flux level in the closed magnetic circuit. These phase shifters

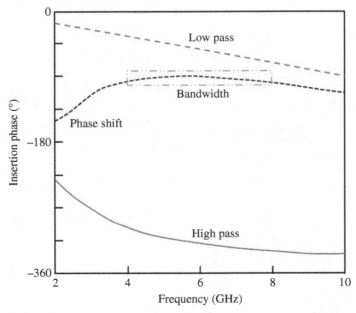

FIGURE 6.26 Insertion phase as a function of frequency for a switched filter phase shifter using a high pass and low pass filter.

change phase by modifying the propagation constant of the ferrite in the transmission line. The rotary-field phase shifter produces has inherently low phase error since the phase shift results from changing the angular orientation of the bias field using a pair of orthogonal windings rather than the magnitude of the bias field.

Table 6.6 lists the characteristics of several types of phase shifters. Ferroelectrics and MEMS dominate the low cost. High-power applications need ferrite phase shifters that also consume a lot of DC power. Ferrites and MEMS have slow switching speeds.

Ferrite phase shifters are too slow for rapid beam scanning. MEMS phase shifters change state in milliseconds but have limited power-handling capability (perhaps

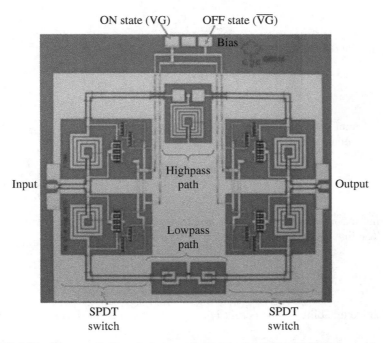

FIGURE 6.27 Diagram (left) and photograph (right) of a GaN HEMT high-pass/low-pass phase shifter. Reprinted by permission of Ref. [29]; © 2012 IEEE.

TABLE 6.6 Phase Shifter Characteristics[a]

Characteristic	Ferroelectric	Ferrite	Semiconductor/MMIC	MEMS
Cost	Low	Very high	High	Low
Reliability	Good	Excellent	Very good	Good
Max power handling	>1 W	kW	>10 W	<50 mW
Switch speed	ns	10–100 μs	ns	10–100 μs
DC power consumption	Low	High	low	Negligible
Size	Small	Large	Small	Small

[a]Ref. [31].

100 mW) and need expensive packaging to protect the movable MEMS bridges against the environment. MMIC phase shifters are fast and can easily change state in tens of nanoseconds, but power handling is usually in the milliwatts but can be higher. They can also be expensive, as they are processed on GaAs, not Si. PIN diode phase shifters are very low loss, but they are controlled by current, not voltage.

Newer phase shifter technology includes tunable dielectric phase shifters made by printing traces on tunable dielectric substrates (Fig. 6.28). Applying a voltage V_{min} results in a substrate relative dielectric constant of ε_{min}, while a voltage V_{max} results in a substrate relative dielectric constant of ε_{max}. The wave velocity when V_{min} is applied is

$$v_1 = \frac{c}{\sqrt{\varepsilon_{min}}} \tag{6.17}$$

Changing the bias to V_{max} results in

$$v_2 = \frac{c}{\sqrt{\varepsilon_{max}}} \tag{6.18}$$

The maximum phase shift experienced by the signal over a distance associated with the two extremes of dielectric constants:

$$\delta = 2\pi f \ell \left(\frac{1}{v_2} - \frac{1}{v_1} \right) = k\ell \left(\sqrt{\varepsilon_{max}} - \sqrt{\varepsilon_{min}} \right) \tag{6.19}$$

Note that the phase shift is a linear function of frequency, so it is actually a time delay. The line length needed to get a 360° phase shift is

$$\ell = \frac{\lambda}{\sqrt{\varepsilon_{max}} - \sqrt{\varepsilon_{min}}} \tag{6.20}$$

The relative tunability of the permittivity of a system is defined as follows [23]:

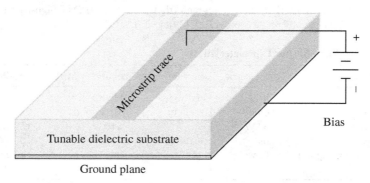

FIGURE 6.28   Microstrip rendition of a tunable dielectric phase shifter.

$$n_r = \frac{\varepsilon_{max} - \varepsilon_{min}}{\varepsilon_{max}} \qquad (6.21)$$

Tunabilities as high as 75% have been reported for (Ba,Sr)TiO$_3$ thin films at 1500 kv/cm [31]. A high electric field generates a large tunability, but the tuning voltage is low (<25 V) because the films are in the nanometer thickness range. An FOM for tunable dielectric devices that includes the device loss is [31]

$$FOM_e = \frac{Q(V_{min})Q(V_{max})(n_r - 1)^2}{n_r} \qquad (6.22)$$

where Q is the inverse of the device loss, $Q(V_{min})$ and $Q(V_{max})$ are at 0 V bias and at the maximum voltages. *FOM* factors as high as 500 have been reported for paraelectric SrTiO$_3$ thin films on SrRuO$_3$ conductors [32].

The dielectric permittivities of thin films and bulk materials change under an applied electric field. Epitaxial paraelectric Ba$_{0.5}$Sr$_{0.5}$TiO$_3$ (BST) thin films deposited on single crystal LaAlO$_3$ and MgO substrates have large changes in permittivity under a DC voltage bias [33]. The loss of the dielectric film is the Q value averaged between the zero field and high electric field bias. Bulk BST ceramics have high tunability as well as higher Q values that BST films. Recent results have shown that high electric fields can be applied to bulk ceramics with nanometer sized grains, corresponding to higher high overall tunability [34].

Discrete liquid crystal phase shifter technology has been recently developed and tested in three different versions [35]: (1) an inverted microstrip line on a glass substrate, (2) a low temperature co-fired ceramic (LTCC) integrated phase shifter using an inverted microstrip line, and (3) a waveguide with liquid crystal making up 1/3 of the cross-sectional area. These phase shifters were shown to be low loss and have switching speeds of many milliseconds. There has been some research on the dielectric constant and group delay of liquid crystals at lower frequencies [36] that suggests liquid crystals will work as phase shifters/time delay units at low microwave frequencies as well [37].

A distributed phase shifter is a transmission line with periodically placed tunable parallel plate capacitors [38]. Capacitance loading creates a periodic structure, with a pass-band that must contain the frequencies of interest [39]. The distributed impedance is given by

$$Z_d = \sqrt{\frac{L_0}{C_0 + C_d}} \qquad (6.23)$$

where L_0 = transmission line inductane, C_0 = transmission line capacitance, and C_d = tunable distributed capacitance

The corresponding phase velocity is given by

$$v = \frac{1}{\sqrt{L_0(C_0 + C_d)}} \qquad (6.24)$$

L_0 and C_0 are functions of the transmission line geometry and material properties. Adding the distributed capacitance lowers the effective characteristic impedance which means that the intrinsic impedance of the transmission line should be larger than the desired matching impedance. The periodicity of the tunable capacitors cause small reflections that add in phase at a certain frequency known as the Bragg frequency:

$$f_{\text{Bragg}} = \frac{1}{\pi \ell \sqrt{L_0 (C_0 + C_{\text{d}})}} \tag{6.25}$$

where ℓ is the constant spacing between the tuning capacitors. The phase shifter operates significantly below f_{Bragg}. The Bragg frequency occurs when the spacing between the capacitors is 1/4 of a wavelength on the transmission line.

$$\delta = 2\pi f \ell \sqrt{L_0} \left(\sqrt{C_0 + C_{\text{max}}} - \sqrt{C_0 + C_{\text{min}}} \right) \tag{6.26}$$

where C_{min} and C_{max} denote the minimum and maximum tuning capacitances. Cascading a sufficient number of sections produces the desired differential phase shift. Increasing ℓ brings the Bragg frequency closer to the operating frequency and reduces the number of required sections to get the desired phase shift. Increasing Z_0 lowers C_0 and reduces the number of sections as well.

6.6   ATTENUATORS

Attenuators are passive devices that reduce the magnitude of a signal passing through it. The most common forms (Fig. 6.29) have two different resistance values that are derived from the desired attenuation in dB, αdB. If the ratio between the input and output voltages is $C = 10^{\alpha\text{dB}/20}$, then the resistor values are [40]

$$R_1 = Z_{\text{c}} \left(\frac{C-1}{C+1} \right) \quad R_3 = Z_{\text{c}} \left(\frac{C^2 - 1}{2C} \right)$$

$$R_2 = Z_{\text{c}} \left(\frac{2C}{C^2 - 1} \right) \quad R_4 = Z_{\text{c}} \left(\frac{C+1}{C-1} \right) \tag{6.27}$$

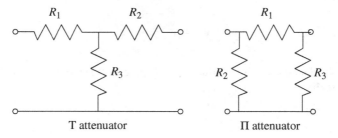

FIGURE 6.29   Two common types of RF attenuators.

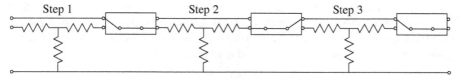

FIGURE 6.30   Diagram of a stepped attenuator.

where $Z_c Z_c$ is the input impedance of the attenuator and the transmission line. Attenuators may be cascaded to get higher values of attenuation. The resulting attenuation is the sum (in decibel) of the attenuation of all the sections. Digitally controlled attenuators, called stepped attenuators, switch fixed attenuation in and out of the signal path as shown in Figure 6.30. An attenuator should not be placed between an element and a low-noise amplifier if noise figure is important, since every dB of attenuation raises the noise figure by 1 dB [40]. A temperature compensating attenuator compensates for gain variations as a function of frequency.

6.7   LIMITER

A limiter is a diode with a low resistance above a threshold voltage, V_T. The diode connects the input conductor to the ground plane. RF signals below V_T pass while those above are rectified to V_T. Si diode-based hybrid limiters are often used, because they are better thermal conductors than GaAs. Limiters are insensitive to parasitic variations and incompatible with GaAs amplifier planar processing, and therefore, are typically not integrated on the same wafer as the LNA [1].

6.8   CIRCULATOR

A circulator is a wideband (larger than an octave) 3-port device that passively connects the transmitter to the antenna while isolating the receiver, or it connects the antenna to the receiver while isolating the transmitter [16]. A circulator can also be an isolator if one port is terminated in a matched load. Circulators are usually much larger than switches and are built using different manufacturing processes than MMICs or hybrids. Also, a switch does not isolate the transmitter from the element impedance variation as the array is scanned, causing the power amplifier performance to be a function of frequency [1]. On the other hand, switches fail to isolate the transmitter from large signals that may enter the antenna.

6.9   CORRECTING ERRORS THROUGH CALIBRATION
AND COMPENSATION

Thermal noise in the array components, elements, and feed network induce random amplitude and phase errors in the signals. In addition, component tolerances, frequency dependence, and beam scanning add uncertainty to the signals. Unexpected failures, particularly with mechanical components, introduce other errors. All of these errors raise the sidelobe levels, cause beam pointing errors, and lower gain. If the errors are random, then they are statistically independent from element to element. Three types of random errors considered here are as follows:

1. Random amplitude error, δ_n^a
2. Random phase error, δ_n^p
3. Random element failure, $P_n^e = \begin{cases} 1 & \text{element functioning properly} \\ 0 & \text{element failure} \end{cases}$

The rms sidelobe level of the array factor assuming random amplitude and phase errors with random element failures is [41]

$$\text{sll}_{\text{rms}} = \frac{(1-P_e)+\overline{\delta_n^{a^2}}+P_e\overline{\delta_n^{p^2}}}{P_e\left(1-\overline{\delta_n^{p^2}}\right)\eta_t N} \tag{6.28}$$

Adding elements to the aperture reduces the rms sidelobe level for the same error levels.

If the same error is common to groups of elements, then that error is correlated within that group of elements even though the error source is random. An example would be a random amplitude error at the subarray port that affects all the elements in that subarray. That random error appears at each element in the subarray, so it is the same for all the elements of that subarray, hence a correlated error shared by elements of that subarray. Central amplification distributes some random errors to all elements, so those errors are correlated at the elements in the same subarray. If the difference between the desired and quantized amplitude weights is a uniformly distributed random number with the bounds being the maximum amplitude error of $\pm\Delta_a/2$, then the rms amplitude error is $\delta_n^a = \Delta_a/\sqrt{12}$ [5]. Substituting this value into (6.28) generates an estimate of the rms sidelobe level. If no two adjacent elements receive the same quantized phase shift, then the difference between the desired and quantized phase shifts appear to be uniform random variables between $\pm\Delta_p/2$. As with the amplitude error, the random phase error formula in this case is $\delta_n^p = \Delta_p/\sqrt{12}$. These same quantized phase errors result in beam pointing errors too.

When two or more elements have the same quantized phase shift, then the error is correlated and quantization lobes form. This situation occurs when the beam is steered to a small angle off boresight. The grating lobes due to these element groupings occur at [5]

$$u_m = u_s \pm \frac{m\lambda}{N_Q d_e} = u_s\left[1\pm\frac{m(N-1)2^{N_{bp}}}{N}\right] \simeq u_s\left(1\pm m2^{N_{bp}}\right) \tag{6.29}$$

T/R module calibration insures desired performance at each element [42]. Arrays are calibrated to remove signal uncertainty by adjusting the attenuators and phase shifters in the T/R modules. Calibration starts by tuning the attenuators at each element to get the desired amplitude. The attenuator's insertion phase can vary as a function of the phase setting. Next, the phase shifters are adjusted to compensate for insertion phase errors. This calibration is done across the bandwidth, range of operating temperatures, and scan angles. If the phase shifter's gain varies as a function of setting, then the attenuators need compensation as well. The process of adjusting the phase and amplitude iterates until the error is minimized. These calibration settings are assembled in a calibration table that is stored in memory. A controller accesses the calibration table and appropriately adjusts the attenuators and phase shifters for the appropriate frequency, temperature, and scan angle.

REFERENCES

[1] B. A. Kopp *et al.*, "Transmit/receive packaging: electrical design issues," *Johns Hopkins APL Tech. Dig.*, vol. 20, pp. 70–80, 1999.

[2] N. J. Kolias and M. T. Borkowski, "The development of T/R modules for radar applications," IEEE MTT-S International Symposium, Montreal, Quebec, Canada, June 2012.

[3] M. Harris *et al.*, "GaN-based components for transmit/receive modules in active electronically scanned arrays," CS MANTECH Conference, New Orleans, LA, USA, May 2013.

[4] D. Vye *et al.*, "The new power brokers: high voltage RF devices," *Microw. J.*, vol. 52, pp. 22–44, 2009.

[5] R. L. Haupt, *Antenna Arrays: A Computational Approach*. Hoboken, NJ: John Wiley & Sons, Inc., 2010.

[6] D. N. McQuiddy *et al.*, "Monolithic microwave integrated circuits: an historical perspective," *IEEE Trans. Microw. Theory Tech.*, vol. 32, pp. 997–1008, 1984.

[7] *F-22 raptor avionics* [online]. Available: http://www.globalsecurity.org/military/systems/aircraft/f-22-avionics.htm. Accessed January 29, 2014.

[8] R. H. Medina *et al.*, "T/R module for CASA phase-tilt radar antenna array," Microwave Conference (EuMC), 2012 42nd European, Amsterdam, Netherlands, October 29–November 1, 2012, pp. 1293–1296.

[9] D. A. Bajgot *et al.*, "Surface mount, 50 watt peak GaN power amplifier for low-cost S-band radar," IEEE Phased Array Symposium, Waltham, MA, USA, October 2013, pp. 59–63.

[10] *Technology today highlighting Raytheon's technology* [online], vol. 3, no. 1, pp. 19–23, 2004. Available: http://www.raytheon.com/newsroom/technology_today/archive/2004_Issue1.pdf. Accessed January 7, 2015.

[11] L. Baggen *et al.*, "Phased array technology by IMST: a comprehensive overview," IEEE Phased Array Symposium, Waltham, MA, USA, October 2013.

[12] A. Yadav *et al.*, "Linearization of Saleh, Ghorbani, and Rapp amplifiers with Doherty technique," *SASTech. J.*, vol. 9, pp. 79–86, 2010.

[13] C. Rapp, "Effects of HPA-nonlinearity on a 4-DPSK/OFDM-signal for a digital sound broadcasting system," Second European Conference on Satellite Communications, Liege, Belgium, October 22–24, 1991, pp. 179–184.

[14] K. G. Gard *et al.*, "Characterization of spectral regrowth in microwave amplifiers based on the nonlinear transformation of a complex Gaussian process," *IEEE Trans. Microw. Theory Tech.*, vol. 47, pp. 1059–1069, 1999.

[15] D. Cerovecki and K. Malaric, "Microwave amplifier figure of merit measurement," *Meas. Sci. Rev.*, vol. 8, no. 4, pp. 104–107, 2008.

[16] R. Sorrentino and G. Bianchi, *Microwave and RF Engineering*. Chichester: John Wiley & Sons, Ltd, 2010.

[17] A. Dao, "Integrated LNA and mixer basics," National Semiconductor Application Note 884, April 1993.

[18] C. Hemmi, "Pattern characteristics of harmonic and intermodulation products in broadband active transmit arrays," *IEEE Trans. Antennas Propag.*, vol. 50, pp. 858–865, 2002.

[19] W. A. Sandrin, "Spatial distribution of intermodulation products in active phased arrays," *IEEE Trans. Antennas Propag.*, vol. 21, pp. 864–868, 1973.

[20] M. D. Weiss and R. L. Haupt, "The impact of time delay verses phase steering on the harmonics radiated by a multibeam wideband active transmit array," ICEAA Conference, Turino, Italy, September 2013.

[21] Agilent Technologies, "Understanding RF/microwave solid state switches and their applications," Literature 5989-7618EN, May 21, 2010.

[22] C. L. Goldsmith *et al.*, "Performance of low-loss RF MEMS capacitive switches," *IEEE Microw. Guided Wave. Lett.*, vol. 8, pp. 269–271, 1998.

[23] R. L. Haupt and M. Lanagan, "Reconfigurable antennas," *IEEE Antennas Propagat. Mag.*, vol. 55, pp. 49–61, 2013.

[24] P. Hindle, "The state of RF and microwave switches," *Microw. J.*, vol. 53, pp. 20–36, 2010.

[25] E. R. Brown, "RF-MEMS switches for reconfigurable integrated circuits," *IEEE Trans. Microw. Theory Tech.*, vol. 46, pp.1868–1880, 1998.

[26] S. J. Gross *et al.*, "Lead-zirconate–titanate-based piezoelectric micromachined switch," *Appl. Phys. Lett.*, vol. 83, pp. 174–176, 2009.

[27] P. D. Grant *et al.*, "A comparison between RF MEMS and semiconductor switches," ICMENS, Banff, Alberta, Canada, August 25–27, 2004, pp. 515–521.

[28] *Phase shifters and I-Q modulators* [online]. Available: http://www.admiral-microwaves.co.uk/pdf/herley/herley-catalogue-phase-shifters.pdf. Accessed December 18, 2014.

[29] K. Hettak *et al.*, "A new type of GaN HEMT based high power high-pass/low-pass phase shifter at X band," IEEE MTT-S International Microwave Symposium, Montreal, Canada, June 2012.

[30] W. E. Hord, "Microwave and millimeter-wave ferrite phase-shifters: state of the art reference," *Microw. J.*, Suppl. 81, 1989.

[31] R. R. Romanofsky, "Array phase shifters: theory and technology," in *Antenna Engineering Handbook* (4th ed.), J. L. Volakis, Ed. New York: McGraw-Hill, 2007, pp. 1–25.

[32] C. Basceri *et al.*, "The dielectric response as a function of temperature and film thickness of fiber-textured (Ba, Sr) TiO thin films grown by chemical vapor deposition," *J. Appl. Phys.*, vol. 82, p. 2497, 1997.

[33] O. G. Vendik *et al.*, "Ferroelectric tuning of planar and bulk microwave devices," *J. Supercond.*, vol. 12, pp. 325–338, 1999.

[34] J. S. Horwitz *et al.*, "The effect of stress on the microwave dielectric properties of $Ba_{0.5}Sr_{0.5}TiO_3$ thin films," *J. Electroceram.*, vol. 4, pp. 357–363, 2000.

[35] S. Kwon *et al.*, "Nonlinear dielectric ceramics and their applications to capacitors and tunable dielectrics," *IEEE Elect. Insul. Mag.*, vol. 27, pp. 43–55, 2011.

[36] S. Strunck *et al.*, "Continuously tunable phase shifters for phased arrays based on liquid crystal technology," IEEE Phased Array Symposium, Waltham, MA, USA, October 2013.

[37] Y. Garbovskiy *et al.*, "Liquid crystal phase shifters at millimeter wave frequencies," *J. Appl. Phys.*, vol. 111, no. 5, p. p054504, 2012.

[38] I. Ruso, *Phase shifters* [online]. Available: http://www.qsl.net/va3iul/Phase_Shifters/Phase_Shifters.pdf. Accessed December 18, 2014.

[39] Z. Kelong *et al.*, "Analysis of a Ka-band phase-shifter using distributed MEMS transmission line structure," *Adv. Intell. Soft. Comput.*, vol. 163, pp. 765–772, 2013.

[40] http://www.microwaves101.com/encyclopedia/attenuators.cfm#intro. Accessed December 18, 2014.

[41] E. Brookner, "Antenna array fundamentals-part 2," in *Practical Phased Array Antenna Systems*, E. Brookner, Ed. Norwood, MA: Artech House, 1991, pp. 3–15.

[42] R. Sorace, "Phased array calibration," *IEEE Trans. Antennas Propag.*, vol. 49, pp. 517–525, 2001.

7

TIME DELAY IN A CORPORATE-FED ARRAY

The distinction between phase shift and time delay becomes very important when considering wideband signals. A single tone represented by a sinusoid having constant time delay and phase shift terms in the argument is shown in Figure 7.1. Phase shift and time delay are constants in the cosine argument. The phase shift is bound by $0 \leq \delta_s < 2\pi$, while the time delay has no bounds. When dealing with only one frequency (narrow band), phase shift and time delay are identical. The phase shift associated with time delay $(2\pi f T_d)$ is a linear function of frequency and has a constant Fourier transform with respect to frequency as shown at the top of Figure 7.2. By contrast, the Fourier transform of a signal time delay is a linear function of frequency as shown at the bottom of Figure 7.2. As mentioned in Chapter 5, time delay is also known as group delay (envelope delay). Group delay is [1]

- a measure of component phase distortion,
- the signal transit time through a component as a function of frequency,
- the negative of the rate of change of phase through a component.

Phased arrays use phase shifters to electronically steer the main beam at the carrier frequency. Narrow band signals can be approximated by the carrier frequency to within reasonable accuracy. In broadband signals, however, the signal envelope has frequency components extending far from the carrier. Large, wideband phased arrays

Timed Arrays: Wideband and Time Varying Antenna Arrays, First Edition. Randy L. Haupt.
© 2015 John Wiley & Sons, Inc. Published 2015 by John Wiley & Sons, Inc.

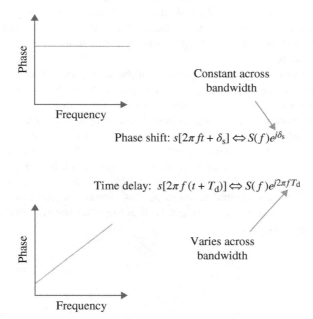

Time delay has units of seconds.

$$\cos\left[2\pi f\left(t + T_d\right) + \delta_s\right]$$

Phase shift has units of radians or degrees

FIGURE 7.1 Time delay and phase shift in the argument of a Fourier harmonic.

Constant across
bandwidth

Phase shift: $s[2\pi ft + \delta_s] \Leftrightarrow S(f)e^{j\delta_s}$

Time delay: $s[2\pi f(t + T_d)] \Leftrightarrow S(f)e^{j2\pi fT_d}$

Varies across
bandwidth

FIGURE 7.2 Phase shift is a constant with respect to frequency. Time delay is a linear function of frequency.

distort signals due to beam squint [2] and pulse dispersion [3]. The phase shift that steers the main beam to (θ_s, ϕ_s) at the center frequency steers the main beam to an offset location at a different frequency. Main beam pointing error that is a function of frequency is known as beam squint. Pulse dispersion occurs when signals do not arrive at all the elements at the same time, because they are incident from an off-broadside angle. Phase shifters align these signals in phase but not in time. As a result, adding the phase-shifted element signals together at the array output causes the pulses to coherently add but also causes pulse spreading in time.

This chapter motivates the need for time delay and explains how to integrate time delay into the array design. Time delay uses some of the same technology as phase shifters, but time delay units are more complex and bigger. Future arrays will rely upon time delay units in order to achieve desired performance standards.

7.1 PULSE DISPERSION

Figure 7.3 shows a narrow band signal (single tone) incident on an array. The signals at all the elements can be aligned by adjusting the phase. In this case, adjusting the phase at each element aligns the phase and results in a coherent addition of the signals and steering the main beam. Aligning all the signals then adding them together produces an output that is NG times the received signal, where N is the number of elements and G is the amplifier gain at each element. Adjusting only the phase at each element does not result in the same signal amplification for wideband signals. Figure 7.4 shows an OOK signal incident on an array. Aligning the phase of the carrier will not necessarily align the envelopes. Only time delay aligns the envelopes at each element, because the total delay needed is much more than one period of the carrier or 360° in phase.

Pulse dispersion is a function of incident angle. Figure 7.5 is an example of a pulse incident on a 4-element array at a small angle off boresight. In this case, phase shifters easily align all four signals, so they are summed to get an amplified pulse with no dispersion. If the signal is incident at a greater angle, then the sum of the pulses spread out in time as shown in Figure 7.6. Using phase shifters to align these signals makes them coherent, but does not align all the envelopes. When the element signals are summed, they produce a pulse that lasts longer than the individual pulse as shown in the bottom of Figure 7.6. The resultant pulse reaches a peak amplitude of 4 but has a rounded envelope that spreads out in time. Time delay in lieu of phase shift aligns the envelopes so that the resulting summed pulse has no distortion as shown in Figure 7.7.

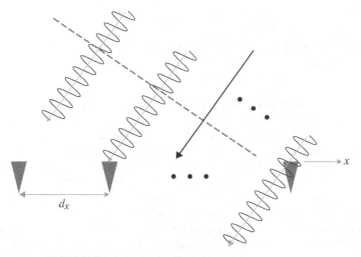

FIGURE 7.3 Narrow band signal incident on an array.

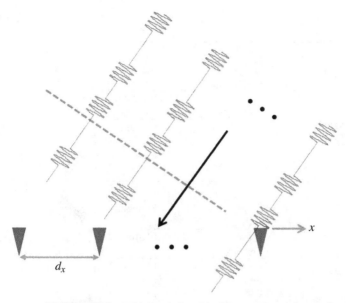

FIGURE 7.4 Wideband signal incident on an array.

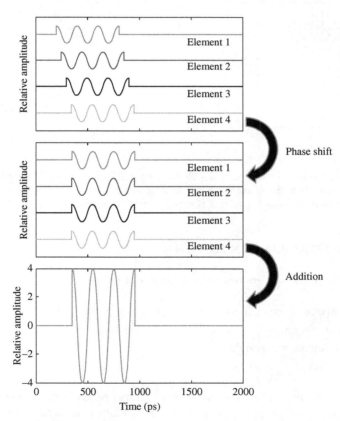

FIGURE 7.5 A signal received by a 4-element array at a small scan angle (top). The envelopes of the signals at the elements can be aligned with phase shift (middle). Coherently adding them together results in an output with no dispersion (bottom).

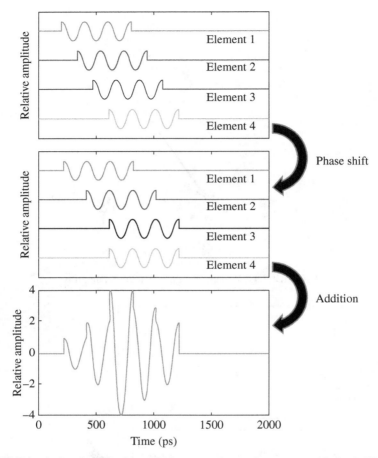

FIGURE 7.6 A signal received by a 4-element array at a large scan angle (top). The envelopes of the signals at the elements cannot be aligned with phased shifters even though the carrier is aligned in phase (middle). Coherently adding them together results in dispersion (bottom).

7.2 PHASED ARRAY BANDWIDTH

The array bandwidth is determined by a number of factors [4]:

1. Bandwidth of the elements in the array
2. Element spacing
3. Maximum steering angle
4. Array size

The element bandwidth needs to encompass the array bandwidth. Grating lobes usually limit the highest frequency in the array bandwidth and are a function of element

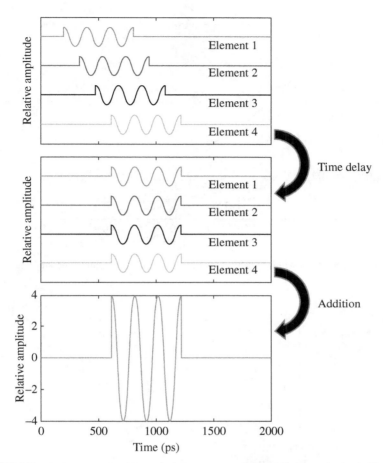

FIGURE 7.7 A signal received by a 4-element array at a large scan angle (top). The envelopes of the signals at the elements can be aligned with time delay (middle). Coherently adding them together results in an output with no dispersion (bottom).

spacing and maximum scan angle. The array size also limits the upper frequency due to pulse dispersion.

The phased array bandwidth can be approximated by twice the 3 dB beamwidth [5]

$$BW\,(\%) = 2\theta_{3\,dB}\ \text{(degrees)} \tag{7.1}$$

In other words, a large phased array (narrow beamwidth) has a narrow bandwidth. Figure 7.8 is a plot of array bandwidth versus scan angle for several different sizes of linear arrays with $d = 0.5\lambda$. Equation 7.1 does not apply to arrays that use time delay.

If the first element in a uniform linear array receives the leading edge of the pulse and the last element receives the trailing edge (Fig. 7.9), then

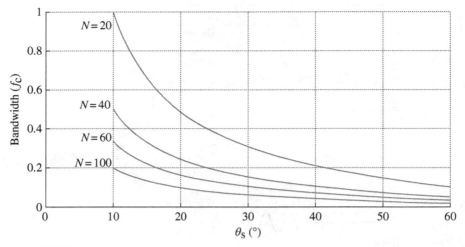

FIGURE 7.8   Bandwidth as a function of the number of elements and steering angle.

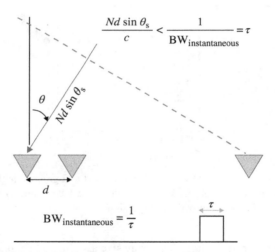

FIGURE 7.9   The array instantaneous bandwidth.

$$Nd \sin \theta_s \leq c\tau \tag{7.2}$$

where τ = pulse width. An approximation to the bandwidth follows from

$$\frac{c}{Nd \sin \theta_s} \geq BW_{instantaneous} = \frac{1}{\tau} \tag{7.3}$$

which is an expression for the time-bandwidth product—a nice result of this analysis.

7.3   TIME DELAY STEERING CALCULATIONS

The first step in specifying a time delay unit is to find the maximum time delay needed to perfectly align the signal envelopes at the maximum scan angle and maximum operating frequency [6]. The maximum time delay across an aperture is

$$\tau_{max} = \frac{D_{max} \sin\theta_s}{c} \qquad (7.4)$$

where D_{max} is the maximum dimension of the aperture. For a rectangular aperture, D_{max} is the diagonal, and for a circular aperture, D_{max} is the diameter. Figure 7.10 shows the difference between the maximum time delay for narrow band and wideband signals. An $N_x \times N_y$ planar array with a rectangular lattice has a maximum time delay given by

$$\tau_{max} = \sin\theta_s \left[N_x d_x \cos\phi_s + N_y d_y \sin\phi_s \right]/c \qquad (7.5)$$

This equation uses N_x and N_y instead of $(N_x - 1)$ and $(N_y - 1)$. The differences are small and their result prevents underestimation of the required time delay. Figure 7.11 is a plot of τ_{max} as a function of N, frequency, and scan. Decreasing the frequency, increasing N, and increasing θ_s causes τ_{max} to increase.

Now that τ_{max} is known, the quantization levels and numbers of bits must be found. A first guess for the most significant bit (MSB) of time delay is approximately half of the maximum time delay.

$$\tau'_{MSB} = \frac{\tau_{max}}{2} \qquad (7.6)$$

For a square array with a square lattice (7.6) becomes

$$\tau'_{MSB} = \frac{0.707 N_x d \sin\theta_s}{c} \qquad (7.7)$$

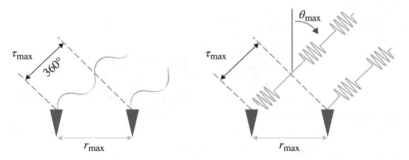

FIGURE 7.10   The maximum time delay for a narrow band (left) and wideband signal (right).

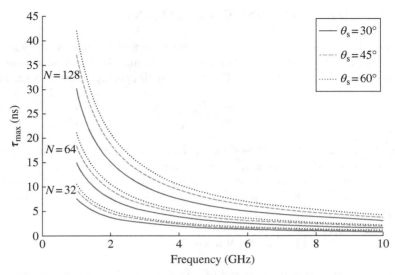

FIGURE 7.11   Maximum time delay versus frequency versus N versus scan angle.

because $N_x = N_y$ and $d = d_x = d_y$. Since the element spacing is a function of θ_s and f_{hi}, then

$$\tau'_{MSB} = \frac{0.707 N_x \sin \theta_s}{(1 + \sin \theta_s) f_{hi}} \qquad (7.8)$$

The MSB of time delay depends on the number of elements, the maximum scan angle, and the highest frequency.

Desired sidelobe levels determines the smallest time delay bit [4]. The root mean square (rms) sidelobe level is relatively constant over all angles and is given by [7]

$$sll_{rms} = \frac{\delta_{LSB}^2}{(1 - \delta_{LSB}^2) \eta_t N} \qquad (7.9)$$

where δ_{LSB} is the least significant bit (LSB) of the phase shifter given by

$$\delta_{LSB} = \frac{2\pi}{2^{N_{bits}} \sqrt{12}} \text{ radians} \qquad (7.10)$$

and η_t is the taper efficiency. This worst case phase LSB then converts to a time delay LSB at the highest frequency.

$$\tau'_{LSB} = \frac{\delta_{LSB}}{2\pi f_{hi}} \qquad (7.11)$$

The number of bits is then found from

$$N_{bits} = ceil\left[\log_2\left(\frac{\tau_{max}}{\tau'_{LSB}} \right) \right] \qquad (7.12)$$

where *ceil* is a function that rounds up to the nearest integer. The time delay unit LSB is based on the number of bits.

$$\tau_{LSB} = \frac{\tau_{max}}{2^{N_{bits}}} \qquad (7.13)$$

A more accurate estimate of the MSB is then

$$\tau_{MSB} = \frac{\tau_{max}}{\left(1 + \frac{1}{2} + \frac{1}{4} + L + \frac{1}{2^{N_{bits}-1}} \right)} \qquad (7.14)$$

7.4   TIME DELAY UNITS

Time delay units use some of the same technology as the phase shifters in Chapter 6. They are not as readily available today as phase shifters, because they are more complex and expensive and have not been needed in most arrays until recently. As the signal bandwidth, array size, and scan volume increase on new array systems, then time delay units become necessary. This section presents some of the technologies used in making time delay units.

The pi network in Figure 7.12 is one way to implement a lumped circuit time delay. Its total time delay is the sum of the time delays for all the pi network stages. If all the pi networks have the same inductors and capacitors, then the time delay is just

$$\tau = N\sqrt{LC} \qquad (7.15)$$

Figure 7.13 has a block diagram of a trombone line time delay unit implemented as a SiGe chip [8]. The 4-bit time delay varies from 4 to 64 ps in increments of 4 ps. It has two sections: a 1-bit 0/32 ps switched delay line and a 3-bit variable trombone delay line. The trombone line extends by turning on the appropriate path-select

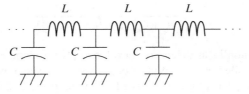

FIGURE 7.12   Lumped circuit time delay.

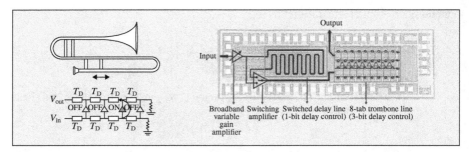

FIGURE 7.13 A 4-bit trombone line time delay architecture (left) with a block diagram superimposed on the chip microphotograph (right). Reprinted by permission of [8]; © 2008 IEEE.

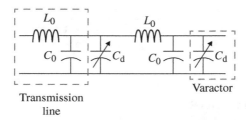

FIGURE 7.14 Model of a varactor loaded transmission line and equivalent lumped circuit.

amplifier that shorts the input line to the output line. This amplifier not only acts like a switch but also boosts the signal amplitude due to losses encountered in the increased path length. In order to save chip area, lumped inductors and capacitors replace lengths of transmission line. The transmission line characteristic impedance and length determine the values of the lumped inductors and capacitors. On-chip inductor loss limits the maximum delay in a practical trombone line. Other limitations include the finite input and output impedances of path select amplifiers that load the lines and the gain and group delay fluctuations over the bandwidth as the number of delay sections is increased. Attenuators in the short paths or amplifiers in the long paths ensure the signal output level is not a function of the time delay.

A distributed phase shifter (Chapter 6) can be converted to a nonlinear delay line. Relatively high impedance transmission line varactors are periodically placed along the time delay line to provide a variable shunt capacitance (Fig. 7.14). The varactor bias controls the time delay via the varactor capacitance that varies between C_{min} and C_{max} [9].

$$\tau = \sqrt{L(C + C_{max})} - \sqrt{L(C + C_{min})} \qquad (7.16)$$

A 1×8, hybrid nonlinear delay was developed [10] to demonstrate up to $18°$ of electronically controlled beam steering from 4 to 5 GHz. The hybrid nonlinear delay lines have up to 267 ps analog time delay (TTD) with less than 5 dB measured insertion loss (Fig. 7.15).

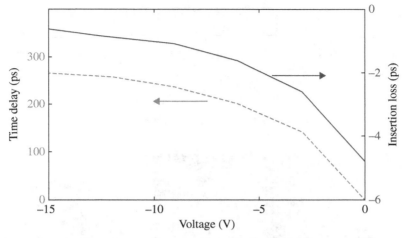

FIGURE 7.15 Measured time delay and insertion loss of a typical hybrid nonlinear delay line [10]

Optical delay lines convert RF signals into light, because optical delay is much shorter than the equivalent RF delay [11]. They resist electromagnetic interference and are low loss, small, and light weight [12]. Two approaches to optical time delay are guided-wave (e.g., fiber) delay lines and free space. Planar optical waveguides [13] are best for very short delays, while fibers are best for long delays [14]. Fiber lengths must be precisely cut. Length errors can be corrected by adding shorter time delay units. Fiber Bragg gratings [15] and dispersive fiber [16] or prisms [17] reuse the same hardware for different elements but require tunable lasers with precisely controlled wavelengths.

Free-space optical time delay has the advantage of having many light beams share the same medium [18]. Long delays, however, require large volumes due to beam expansion. Free-space approaches generally do not need multiple or tunable lasers. In addition, the time delay is nondispersive.

A bulk acoustic wave (BAW) time delay device converts an RF signal into an acoustic wave that is easy to delay in a crystal before reconverting it back to RF [19]. The BAW device in Figure 7.16 accepts an RF signal through a matching network and then sends it to a transducer made of ZnO piezoelectric thin film that converts an electromagnetic wave into an acoustic wave. The metalized electrodes on either side of the ZnO are made from a gold composite metallization. Acoustic waves travel about 1 μs/cm through the delay crystal before being reconverted to an RF signal that is finally matched to the output. BAW delay lines have bandwidths that are a fraction of the 300 MHz to 18 GHz range and provide delays from 100 ns to several milliseconds. Time delay in the millisecond range is difficult at higher frequencies due to losses. Reflection BAW time delays that use only one port are also available.

Time delay units are not as readily available on the commercial market as phase shifters. Cobham Defense Electronics is one of the few companies that has developed a commercial time delay module [20]. The 13 mm × 9 mm plastic packaged module

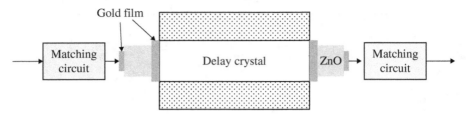

FIGURE 7.16   Diagram of a BAW delay line.

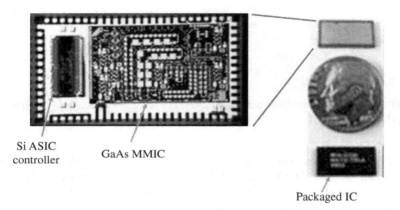

FIGURE 7.17   Cobham TDU/ASIC module. Courtesy of Cobham Defense Electronics.

combines a PHEMT GaAs MMIC and a silicon controller into a QFN quad-flat no-leads (QFN) style package (Fig. 7.17). This time delay unit has over 20 dB gain and operates over 0.8–8 GHz (Fig. 7.18). The 8-bit time delay has an LSB of 4 ps with a total maximum time delay of 764 ps. A 6-bit attenuator on the GaAs chip attenuates the signal from 0 to 31.5 dB in 0.5 dB steps. The module needs ±3.3 V power supplies. It accepts serial commands to adjust the time delay and attenuation. The module has a 7-dBm P1dB, 20-dB gain, 5-dB NF, and 18-dB I/O return loss.

The Cobham time delay chip was used in a timed array capable of a ±60° scan in azimuth (Fig. 7.19). It has 8 columns of 16 elements (256 elements total). The output from each column is combined and sent to the time delay unit board which forms beams (Fig. 7.20). The array has a 2 GHz instantaneous bandwidth and a 1–8 GHz operational bandwidth. Figure 7.21 shows measured patterns at 6, 7, and 8 GHz for broadside and main beam steered to 26°. Unlike phase steering, all the beams point to 26°.

7.5   UNIT CELL CONSTRAINTS

Time delay bits take up space. Large bits need more space than small bits, because they have longer line lengths, more components, and extra amplifiers. All the time delay bits immediately behind the element must fit inside the unit cell [6]. The MSB

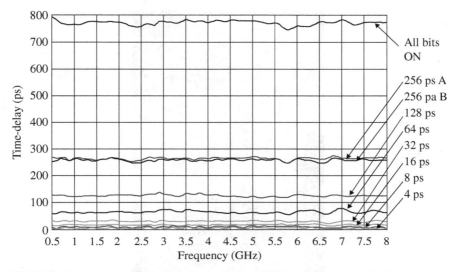

FIGURE 7.18   Measured time delay for the Cobham TDU chip. Courtesy of Cobham Defense Electronics.

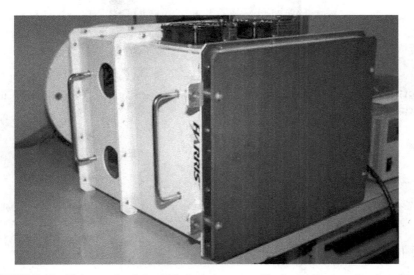

FIGURE 7.19   Photograph of the 256-element array with time delay steering in azimuth. Courtesy of Cobham Defense Electronics.

determines the longest path and should take up less than half the unit cell area. All the other bits must fit in the other half. Bits can be placed on multiple layers and connected by vias. Coupling and path loss are additional problems to overcome within a constrained.

FIGURE 7.20 The 8-channel time delay beamformer. Courtesy of Cobham Defense Electronics.

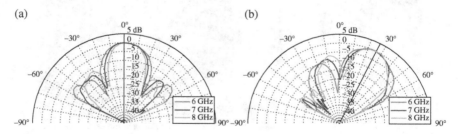

FIGURE 7.21 Measured array patterns: (a) boresight and (b) steered to 26°. Courtesy of Cobham Defense Electronics.

A 32×32 array with a square element lattice that operates from 9 to 11 GHz with a maximum elevation scan of 60° has an element spacing of $d = 1.46$ cm. If the time delay bits in the time delay units are lengths of microstrip line, then all the bits must fit behind the element otherwise, large bits of the time delay are placed at the

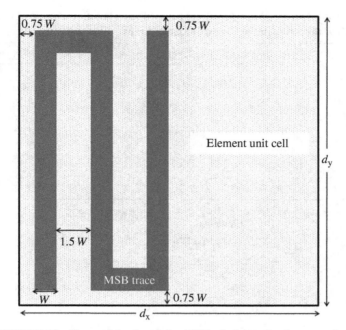

FIGURE 7.22   Trace of the time delay MSB relative to the element unit cell.

subarray level. Figure 7.22 shows how the MSB of the time delay fits within half of the unit cell. The trace width is $W = 1.5$ mm mm for $50\,\Omega$ impedance [21], and the separation between traces is $1.5W$. For a PCB with a relative dielectric constant of $\varepsilon_r = 2.4$ and a substrate thickness of 0.5 mm, the effective dielectric constant is $\varepsilon_{\text{eff}} \approx 2$. The time delay MSB in Figure 7.22 has a trace length bounded by

$$L < 3d_y = 4.38\,\text{cm} \qquad (7.17)$$

The time delay is found from (5.7). As a result, this MSB has a time delay of 0.21 ns. Fitting the MSB into the unit cell becomes more difficult as the bandwidth, number of elements, and maximum scan angle increase.

An LSB phase shifter bit of $22.5°$ translates into $\tau_{\text{LSB}} = 0.0036$ ns. The length of the microstrip line needed for each bit and the corresponding time delay are shown in Table 7.1. Fitting all of these bits into the unit cell along with switches, amplifiers, and so on may not be possible.

7.6   TIME DELAY BIT DISTRIBUTION AT THE SUBARRAY LEVEL

The previous section had an example demonstrating that space limitations may force large bits to be placed at the subarray level where groups of elements share the large bits. This arrangement is cost effective, because large bits cost more than small bits due to their size and associated electronics needed to compensate for path loss.

TABLE 7.1   Microstrip Line Length and Corresponding Time Delay for Each of the 9 Bits in the Example Planar Array

Bit	Microstrip line length (mm)	Time delay (ns)
1 MSB	282	0.9248
2	141	0.4624
3	71	0.2312
4	35	0.1156
5	18	0.0578
6	9	0.0289
7	4	0.0145
8	2	0.0072
9 LSB	1	0.0036

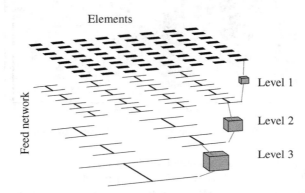

FIGURE 7.23   Placement of time delay at various levels in a corporate feed network.

Figure 7.23 illustrates the size of the time delay package that can be placed at the different subarray levels [22]. Bigger bits require bigger packaging that may only fit at higher subarray levels.

A logical approach to fitting large time delay bits in the array has phase shifters or small time delay bits at the element level and then places large time delay bits at one or more subarray ports. The linear array factor for this strategy when the array has 2^M elements is given by [23]

$$AF = \sum_{n=1}^{N} a_n e^{j\left(kx_n \sin\theta - 2\pi f \sum_{m=1}^{M} T_{dmn} + \delta_n \right)} \tag{7.18}$$

where

$$T_{dmn} = \begin{bmatrix} \tau_{11} & \tau_{12} & \tau_{13} & \tau_{14} & \tau_{15} & \tau_{16} & \cdots & \tau_{1N} \\ \boxed{\tau_{21} \quad \tau_{21}} & \boxed{\tau_{22} \quad \tau_{22}} & \boxed{\tau_{23} \quad \tau_{23}} & & \cdots & & \tau_{2N/2} \\ \vdots & & & & & & \ddots & \vdots \\ \boxed{\tau_{M1} \quad \tau_{M1}} & \boxed{\tau_{M1} \quad \tau_{M1}} & \boxed{\tau_{M1} \quad \tau_{M1}} & & \cdots & & \tau_{M2} \end{bmatrix} \tag{7.19}$$

$$\delta_n = -kx_n \sin\theta_s \qquad (7.20)$$

Shared time delay bits are boxed.

A planar array typically combines four elements or four subarrrays of the same size as shown in Figure 7.24. Equations 7.18–7.20 need to be appropriately modified for the planar array.

$$AF = \sum_{n=1}^{N} a_n e^{j\left(k(x_n \sin\theta \cos\phi + y_n \sin\theta \sin\phi) - 2\pi f \sum_{m=1}^{M} T_{dmn} + \delta_n\right)} \qquad (7.21)$$

where

$$T_{dmn} = \begin{bmatrix} \tau_{11} & \tau_{12} & \tau_{13} & \tau_{14} & \tau_{15} & \tau_{16} & \cdots & \tau_{1N} \\ \tau_{21} & \tau_{21} & \tau_{21} & \tau_{21} & \tau_{22} & \tau_{22} & \cdots & \tau_{2N/4} \\ \vdots & & & & & & \ddots & \vdots \\ \tau_{M1} & \tau_{M1} & \tau_{M1} & \tau_{M1} & \tau_{M1} & \tau_{M1} & \cdots & \tau_{M4} \end{bmatrix} \qquad (7.22)$$

$$\delta_n = -k\left(x_n \sin\theta_s \cos\phi_s + y_n \sin\theta_s \sin\phi_s\right) \qquad (7.23)$$

Figure 7.25 shows the array factors of the isotropic point source planar array in Figure 7.24 when the array is steered to the maximum scan angle of $\left(\theta_s, \phi_s\right) = \left(60°, 45°\right)$ with 4-bit phase shifters at every element. The pattern is a θ-cut at $\phi_s = 45°$. At the

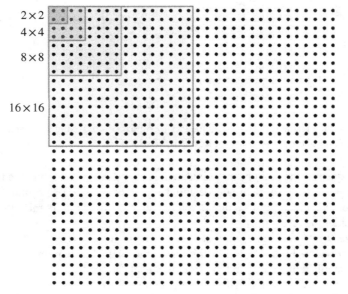

FIGURE 7.24   A 32×32 element array broken into power of two subarrays. Reprinted by permission of Ref. [6] © 2013.

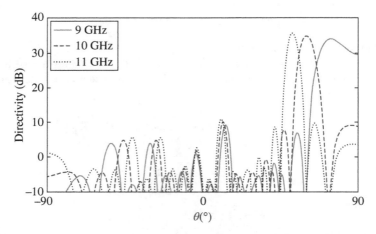

FIGURE 7.25 The main beam is scanned to $(\theta_s,\phi_s) = (60°,45°)$ using phase shifters at the elements. Reprinted by permission of Ref. [6]; © 2013 IEEE.

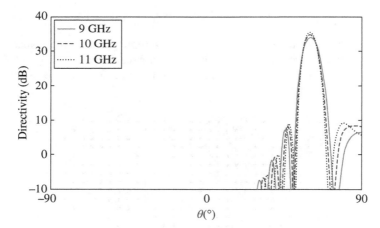

FIGURE 7.26 The main beam is scanned to $(\theta_s,\phi_s) = (60°,45°)$ using time delay units at the elements. Reprinted by permission of Ref. [6]; © 2013 IEEE.

center frequency, the main beam points in the desired direction, while the main beam at 11 GHz squints closer to broadside and the main beam at 9 GHz squints away from broadside. Quantization lobes due to the quantized phase shifters raise the peak and average sidelobe levels [24].

Placing 9-bit time delay units at every element in the array, produces the array factors shown in Figure 7.26. In this case, the main beam points in the desired direction at all frequencies, and the sidelobe levels are well-behaved. Unfortunately, all the time delay bits do not fit within the unit cell, so the large bits must be placed at the subarray levels. Quantization lobes are not noticeable due to the large number of bits.

Comparing (7.17) to the lengths in Table 7.1 reveals that bits 1, 2, and 3 do not fit in the unit cell. These large bits can be placed at the subarray levels where all the elements in the subarray share the large bits. Sharing the large bits with all the elements in the subarray is a cost effective measure if performance is still acceptable.

Our 32×32 element array can be broken into subarrays with 2×2, 4×4, 8×8, and 16×16 subarrays as shown in Figure 7.24. This architecture is ideal for a binary corporate feed network. The power of 2 sizing of the subarrays is also convenient for distribution of the time delay bits which are a power of 2 as well. Quantization lobes become a problem as the size of the subarray increases.

Figure 7.27 shows the steered array factor with 7-bit time delay units at the elements and 2-bit time delay units at the 4×4 subarrays. Compared with Figure 7.26, the main beam still points in the desired direction for all frequencies, but the sidelobes have quantization lobes like Figure 7.25.

Figure 7.28 shows the array factors using 4-bit phase shifters at the elements and 5-bit time delay units at the 4×4 subarrays. The average sidelobe level is higher than either Figure 7.26 or Figure 7.27. There is a very small squint caused by the phase shifters. If 4-bit phase shifters are at the elements and 5-bit time delay units at the 8×8 subarrays, then the sidelobe levels increase, and the beam squint also increases as shown in Figure 7.29. The sidelobes are even higher in this case.

Replacing phase shifters with time delay units will also lower the BER of a communications array. The bottom curve in Figure 7.30 represents the ideal BER versus E_b/N_0 for a 50 Mbps BPSK signal. A 32×32 element array with ideal time delay behind each element has the same BER curve (right above the ideal BER curve) for all scan angles out to $60°$. The remaining curves correspond to various angles when the beam is steered using ideal phase shifters. The highest curve is the worst-case

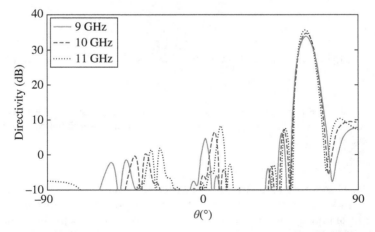

FIGURE 7.27   The main beam is scanned to $(\theta_s, \phi_s) = (60°, 45°)$ using 7-bit time delay units at the elements and 2-bit time delay units at the 4×4 subarrays. Reprinted by permission of Ref. [6]; © 2013 IEEE.

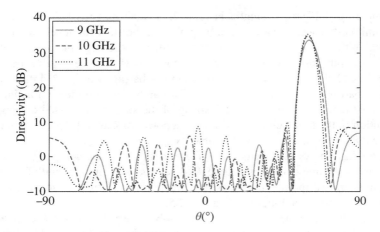

FIGURE 7.28 The main beam is scanned to $(\theta_s, \phi_s) = (60°, 45°)$ using 4-bit phase shifters at the elements and 5-bit time delay units at the 4×4 subarrays. Reprinted by permission of Ref. [6]; © 2013 IEEE.

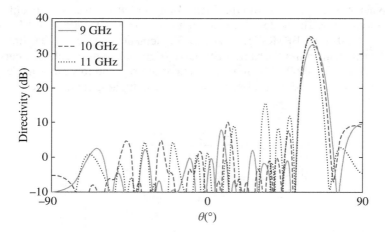

FIGURE 7.29 The main beam is scanned to $(\theta_s, \phi_s) = (60°, 45°)$ using 4-bit phase shifters at the elements and 5-bit time delay units at the 8×8 subarrays. Reprinted by permission of Ref. [6]; © 2013 IEEE.

scan angle. These curves are inferior to the time delay steering curve, because beam squint causes signal distortion. For a particular desired BER, the distance between the outer ideal time delay curve and the highest curve represents the implementation loss for a 50 Mbps signal due to beam steering with phase shifters instead of time delay units. For the example, this implementation loss is approximately 0.35 dB at a BER of 10^{-4}.

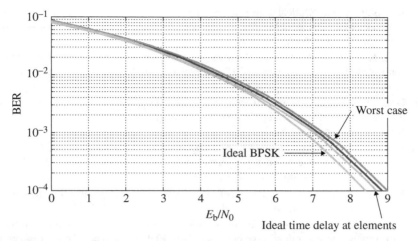

FIGURE 7.30   BER curve at 50 Mbps. Reprinted by permission of Ref. [25]; © 2013 IEEE.

REFERENCES

[1] *Group delay* [Online]. Available http://na.tm.agilent.com/pna/help/WebHelp7_5/Tutorials/Group_Delay6_5.htm. Accessed January 7, 2015.

[2] R. C. Hansen, "Phase and delay in corporate-fed arrays," *IEEE Antennas. Propag Mag.*, vol. 44, pp. 24–29, 2002.

[3] W. W. Shrader, "Universal design curves for effects of dispersion in phased array radars," 15th European Microwave Conference, Paris, France, September 1985, pp. 566–571.

[4] R. L. Haupt, *Antenna Arrays: A Computational Approach*, Hoboken, NJ: John Wiley & Sons, Inc., 2010.

[5] R. J. Mailloux, *Phased Array Antenna Handbook*, 2nd ed., Norwood, MA: Artech House, 2005.

[6] R. L. Haupt, "Fitting time delay units in a large wideband corporate fed array," IEEE Radar Conference, Ottawa, Canada, April 2013.

[7] E. Brookner, editor, "Antenna array fundamentals part 2," in *Practical Phased Array Antenna Systems*, Norwood, MA: Artech House, 1991, pp. 3–8.

[8] H. Hashemi *et al.*, "Integrated true-time-delay-based ultra-wideband array processing," *IEEE Commun. Mag.*, vol. 46, pp. 162–172, 2008.

[9] C. Liang, *True time delay based broadband phased antenna array system* [Online]. Available: http://tempest.das.ucdavis.edu/mmwave/NDL_PAA/NDL1.html. Accessed January 7, 2015.

[10] C. C. Chang, *Nonlinear delay lines* [Online]. Available: http://tempest.das.ucdavis.edu/mmwave/NDL_PAA/. Accessed January 7, 2015.

[11] C. M. Warnky *et al.*, "Demonstration of a quartic cell, a free-space true-time-delay device based on the white cell," *J. Lightw Technol.*, vol. 24, pp. 3849–3855, 2006.

[12] J. Yao, "Microwave photonics," *J. Lightw Technol.*, vol. 27, pp. 314–335, 2009.

[13] L. Eldada, "Laser-fabricated delay lines in GaAs for optically steered phased-array radar," *J. Lightw Technol.*, vol. 13, pp. 2034–2039, 1995.

[14] J-D. Shin *et al.*, "Optical true time delay feeder for X-band phased array antennas composed of 2 × 2 optical MEMS switches and fiber delay lines," *IEEE Photon Technol. Lett.*, vol. 16, pp. 1364–1366, 2004.

[15] D. T. K. Tong and M. C. Wu, "Transmit-receive module of multiwavelength optically controlled phased-array antennas," *IEEE Photon Technol. Lett.*, vol. 10, pp. 1018–1019, 1998.

[16] D. B. Hunter *et al.*, "Demonstration of a continuously variable true-time delay beamformer using a multichannel chirped fiber grating," *IEEE Trans. Microwave Theory Tech.*, vol. 54, pp. 861–867, 2006.

[17] B. Vidal *et al.*, "Novel photonic true-timedelay beamformer based on the free spectral range periodicity of arrayed waveguide gratings and fiber dispersion," *IEEE Photon Technol. Lett.*, vol. 14, pp. 1614–1616, 2002.

[18] Y. Shi and B. L. Anderson, "Robert cell-based optical delay elements for white cell true-time delay devices," *J. Lightw Technol.*, vol. 31, pp. 1006–1014, 2013.

[19] Teledyne Microwave, *BAW delay lines*, OFOISR App 06-S-1942 [Online]. Available: http://www.teledynemicrowave.com/docs/All_PDFs/Teledyne%20Microwave%20BAW%20Products%200507.pdf. Accessed December 17, 2014.

[20] T.W. Dalrymple, *TELA wideband array demonstration* [Online]. Available: http://www.microwaves101.com/downloads/03%20Thomas.Dalrymple.pdf. Accessed April 6, 2010.

[21] *Microstrip Calculator* [Online]. Available: http://www.microwaves101.com/encyclopedia/calmstrip.cfm. Accessed October 24, 2012.

[22] R. L. Haupt, "Tradeoffs in the placement of time delay in a large, wideband antenna array," 8th European Conference on Antennas & Propagation, The Hague, Netherlands, April 2014.

[23] R. L. Haupt, "Time delay unit quantization at the subarray level for large, wideband linear arrays," EMTS Symposium, Hiroshima, Japan, May 2013.

[24] J. Corbin and R. L. Howard, "TDU quantization error impact on wideband phased-array performance," IEEE Phased Array Systems and Technology Conference, Boston, MA, USA, October 2000, pp. 457–460.

[25] P. Moosbrugger *et al.*, "Degradation in theoretical phase shift keying waveforms due to signal dispersion in a large communications phased array," IEEE Phased Array Symposium, Boston, MA, USA, October 2013.

8

ADAPTIVE ARRAYS

An adaptive array improves signal detection by modifying its antenna pattern characteristics in response to environmental or operational changes [1]. Adaptive arrays are timed arrays, since the array performance changes with time. This chapter begins with the development of the signal correlation matrix which is at the heart of direction finding and adaptive nulling algorithms. Since these algorithms primarily rely upon complex and expensive digital beamforming (DBF), some other cheaper alternative adaptive nulling approaches are presented. Next, reconfigurable arrays are introduced. They use mechanical and electrical switching mechanisms to remake the array and alter its performance in a desirable way. Finally, the chapter ends with an overview of several other adaptive arrays that prove useful in various systems.

Adaptive arrays originated in the 1950s with the invention of the Van Atta retrodirective array [2]. Shortly afterward, Howells proposed a method of steering a null in the direction of a jammer [3] in order to improve signal reception. Howells and Applebaum successfully tested a five-loop sidelobe cancelling array, but the information was not reported for many years [4]. The next major contribution was the least mean square (LMS) algorithm, and it caught on very quickly [5]. This approach became the canonical adaptive algorithm from which many variations have sprouted.

Adaptive arrays sense the presence of interference then place nulls in the directions of the interfering signals while maintaining sufficient gain in the direction of the desired signal. Figure 8.1 shows the array factor before and after adaptation with interference incident on a sidelobe. The rest of this chapter explains the algorithms and array architectures of adaptive arrays.

Timed Arrays: Wideband and Time Varying Antenna Arrays, First Edition. Randy L. Haupt.
© 2015 John Wiley & Sons, Inc. Published 2015 by John Wiley & Sons, Inc.

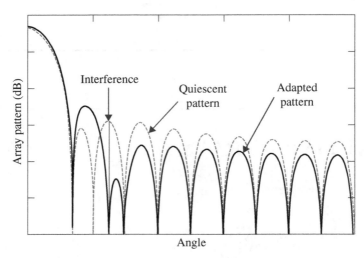

FIGURE 8.1 Adaptive nulling places a null in the sidelobe where interference is present.

8.1 SIGNAL CORRELATION MATRIX

Most adaptive arrays use the signal correlation matrix to characterize signals received by the array. The signal correlation matrix quantifies the relationship between the signals at all the elements in the array. Signals incident on the array are correlated from element to element except for a phase difference due to the different travel times needed to reach the elements, while noise, for the most part, is uncorrelated across elements. These characteristics differentiate signals from noise, so the signals can be located in angle. In order to form the signal correlation matrix, the array must have a receiver or ADC at each element and a means to perform correlation processing.

If M signals arrive at each of the N elements of a planar array, then the vector representing the voltage induced by the mth signal at elements located at positions x_n, y_n is

$$s_m(t)\left[e^{jk(x_1 u_m + y_1 v_m)} \quad e^{jk(x_2 u_m + y_2 v_m)} \quad \cdots \quad e^{jk(x_N u_m + y_N v_m)} \right] \tag{8.1}$$

The array beamformer weights and combines these signals to get the received signal:

$$F(t) = \mathbf{w}^* \mathbf{X}(t, \theta) \tag{8.2}$$

where

$$\mathbf{X}(t, \theta) = \mathbf{A}(\theta)\mathbf{s}^T(t) + \mathbf{N}^T(t) = \text{received signal vector}$$
$$\mathbf{w} = [w_1 \quad w_2 \quad \cdots \quad w_N] = \text{element weight vector}$$

$\mathbf{s}(t) = [s_1(t) \quad s_2(t) \quad \cdots \quad s_M(t)] = \text{transmitted signal vector}$

$$\mathbf{A}_{nm} = \begin{bmatrix} e^{jk(x_1 u_1 + y_1 v_1)} & e^{jk(x_1 u_2 + y_1 v_2)} & \cdots & e^{jk(x_1 u_M + y_1 v_M)} \\ e^{jk(x_2 u_1 + y_2 v_1)} & e^{jk(x_2 u_2 + y_2 v_2)} & \cdots & e^{jk(x_2 u_M + y_2 v_M)} \\ \vdots & \vdots & \vdots & \vdots \\ e^{jk(x_N u_1 + y_N v_1)} & e^{jk(x_N u_2 + y_N v_2)} & \cdots & e^{jk(x_N u_M + y_N v_M)} \end{bmatrix} = \text{array steering vector}$$

$\mathbf{N}(t) = N \times 1$ noise vector

$u_m = \sin\theta_m \cos\phi_m$, $v_m = \sin\theta_m \sin\phi_m$

$^T = $ transpose

$^* = $ complex conjugate

The total array output power is the magnitude squared of the received signal

$$P = 0.5|F|^2 = 0.5E\left\{\left|\mathbf{w}^*\mathbf{X}(t,\theta)\right|^2\right\} = 0.5E\left\{\mathbf{w}^*\mathbf{X}(t,\theta)\mathbf{X}^\dagger(t,\theta)\mathbf{w}^T\right\} = 0.5\mathbf{w}\mathbb{C}\mathbf{w}^\dagger \quad (8.3)$$

where

$\mathbb{C} = E\left\{\mathbf{X}\mathbf{X}^\dagger\right\} = $ covariance matrix

$\dagger = $ complex conjugate transpose of the vector

$E\{\ \} = $ expected value

Assume the signals are zero mean, stationary processes, so that the covariance matrix equals the correlation matrix. The signal correlation separates into four parts

$$\mathbb{C} = E\left\{\mathbf{X}\mathbf{X}^\dagger\right\} + E\left\{\mathbf{N}\mathbf{X}^\dagger\right\} + E\left\{\mathbf{X}\mathbf{N}^\dagger\right\} + E\left\{\mathbf{N}\mathbf{N}^\dagger\right\} \quad (8.4)$$

$$= \mathbb{C}_s + \mathbb{C}_{\text{noise}-s} + \mathbb{C}_{s-\text{noise}} + \mathbb{C}_{\text{noise}}$$

The middle two terms in (8.4) average to zero when the signal and noise are uncorrelated, leaving only the signal and noise correlation matrixes

$$\mathbb{C} = \mathbb{C}_s + \mathbb{C}_{\text{noise}} \quad (8.5)$$

The noise is uncorrelated from element to element, so the diagonal elements of the noise correlation matrix equal the noise variance (σ_{noise}^2)

$$\mathbb{C}_{\text{noise}} = \begin{bmatrix} \sigma_{\text{noise}}^2 & 0 & 0 \\ 0 & \ddots & 0 \\ 0 & 0 & \sigma_{\text{noise}}^2 \end{bmatrix} \quad (8.6)$$

The correlation matrix eigenvectors and eigenvalues correspond to the received signals and noise. The correlation matrix can be decomposed into

$$\mathbb{C}_{\text{noise}} = \mathbf{W}_\lambda \begin{bmatrix} p_1 & 0 & \cdots & 0 \\ 0 & p_2 & \ddots & \vdots \\ \vdots & \ddots & \ddots & 0 \\ 0 & \cdots & 0 & p_N \end{bmatrix} \mathbf{W}_\lambda^{-1} \tag{8.7}$$

where

$$\mathbf{W}_\lambda = \begin{bmatrix} w_{11} & w_{12} & \cdots & w_{1N} \\ w_{21} & w_{22} & & \vdots \\ \vdots & & \ddots & \\ w_{N1} & \cdots & & w_{NN} \end{bmatrix} = \text{matrix whose columns are array weights (eigenvectors)}$$

p_n = eigenvalue associated with column n of $\mathbf{W}_\lambda$

When M uncorrelated signals are incident on an N element array $(N > M)$, they produce M signal eigenvalues and $N - M$ noise eigenvalues. Eigenvalues are proportional to the relative signal and noise powers. Large eigenvalues correspond to signals that are coherent from element to element, while small eigenvalues correspond to the noise. A signal eigenvalue is a function of the signal amplitude and direction, while noise eigenvalues equal the noise variance, $p_{\text{noise}} = \sigma_{\text{noise}}^2$ and are angle independent. Signal eigenvectors are array weights that steer "eigenbeams" in the directions of the received signals [6], ensuring signal reception. Noise eigenvectors, on the other hand, are array weights that steer nulls in the directions of the received signals. Weights may be chosen according to some criteria, most usually so that they optimize the output SINR. This is demonstrated in Section 8.2.

8.2   OPTIMUM ARRAY WEIGHTS

Assume that the array output consists of the desired signal plus and noise. For convenience, the desired signal is the first signal, S_1, in the signal vector in (8.1). The difference between the array signal output and the desired signal is an error signal.

$$\varepsilon = s_1 - \mathbf{w}^* \mathbf{X} \tag{8.8}$$

The mean square error is the expected value of (8.8) squared

$$E\{\varepsilon^2\} = E\{|s_1|^2\} + \mathbf{w}\mathbb{C}\mathbf{w}^\dagger - 2\mathbf{w}^*\mathbf{q}^{\mathsf{T}} \tag{8.9}$$

where $\mathbf{q} = E\{s_1 \mathbf{X}\}$ is the signal correlation vector. This vector quantifies how much the received signal looks like the desired signal. From (8.1), s_1 is the first (desired) signal in the signal vector, $\mathbf{s}(t)$. $E\{\varepsilon^2\}$ is a bowl-shaped surface with a minimum that can be easily found by taking the gradient. Since ε is a function of time, ε^2 is not a constant surface, so taking the gradient does not necessarily lead to the minimum.

Using the instantaneous value rather than expected value works well when ε^2 varies slowly enough that the gradient leads close to the bottom. Taking the gradient of the mean square error yields

$$\nabla_w^T E\{\varepsilon^2\} = 2\mathbb{C}\mathbf{w}^T - 2\mathbf{q}^T \tag{8.10}$$

Setting the gradient equal to zero

$$2\mathbb{C}\mathbf{w}_{\text{opt}}^T - 2\mathbf{q}^T = 0 \tag{8.11}$$

leads to the mean square error optimum weights known as the Wiener–Hopf solution [7]:

$$\mathbf{w}_{\text{opt}}^T = \mathbb{C}^{-1}\mathbf{q}^T = \mathbb{C}^{-1}E[s_1\mathbf{s}^T] \tag{8.12}$$

A direct implementation of (8.12) is not practical for several reasons. First, the desired signal must be known. In practice, an estimate of the desired signal is used, which implies some characteristics of the signal must be known *a priori*. Second, the correlation matrix must be known. Again, an estimate of the correlation matrix is usually used. Another less obvious implementation problem is that the amplitudes and phases of the signals at each element must be known in order to form *C* and **q**. Only DBF yields the required information, incurring the cost and complexity of digital hardware implementation and the required intensive calibration.

The direct matrix inversion (DMI) algorithm, also known as the sample matrix inversion (SMI) algorithm, estimates *C* and finds an approximate w_{opt} using the Wiener–Hopf equation (8.12) [8]. Estimates of the correlation matrix and vector using *K* samples are given by

$$\tilde{\mathbb{C}}(\kappa) = \frac{1}{K}\sum_{\kappa=1}^{K}\mathbf{X}(\kappa)\mathbf{X}^\dagger(\kappa) \tag{8.13}$$

$$\mathbf{q}(\kappa) = \frac{1}{K}\sum_{\kappa=1}^{K}s_1^{\,\dagger}(\kappa)\mathbf{X}(\kappa) \tag{8.14}$$

This sample correlation matrix is inverted and the weights of the *k*th time sample are found from

$$\mathbf{w}(\kappa) = \tilde{\mathbb{C}}^{-1}(\kappa)\mathbf{q}(\kappa) \tag{8.15}$$

An improved estimate of *C* is possible using the recursive least squares (RLS) algorithm [9]

$$\tilde{\mathbb{C}}(\kappa) = \mathbf{X}\mathbf{X}^\dagger(\kappa) + \alpha\,\tilde{\mathbb{C}}(\kappa-1) \tag{8.16}$$

$$\tilde{\mathbb{C}}(\kappa) = x(\kappa)x^\dagger(\kappa) + \alpha\,\tilde{\mathbb{C}}(\kappa-1) \tag{8.17}$$

and the correlation vector estimate is

$$\mathbf{q}(\kappa) = s_1^{\dagger}(\kappa)\mathbf{s}(\kappa) + \alpha\,\mathbf{q}(\kappa - 1) \tag{8.18}$$

where $0 \le \alpha \le 1$. The SINR is a good criterion to determine when the algorithm converges [10]

$$SINR = \frac{\mathbf{w}^{\dagger}\mathbb{C}_s\mathbf{w}}{\mathbf{w}^{\dagger}(\mathbb{C}_i + \mathbb{C}_n)\mathbf{w}} \tag{8.19}$$

An alternative formulation called the minimum variance distortionless response (MVDR) relies upon knowing the signal direction, rather than the desired signal, to minimize the output noise variance. This approach also maximizes the SINR. The optimum weights are given by [6]

$$\mathbf{w}_{\mathrm{opt}}^{\mathrm{T}} = \mathbb{C}_{\mathrm{noise}}^{-1}\mathbf{A}_{n1} \tag{8.20}$$

where $\mathbf{A}_{n1}$ = main beam steering direction.

8.3   ADAPTIVE WEIGHTS WITHOUT INVERTING THE CORRELATION MATRIX

Historically, the first adaptive arrays used the simpler and closely related Howells–Applebaum and LMS algorithms. Where the Weiner–Hopf approach collects and processes blocks of data, these algorithms update the weight vector at every time step and iteratively converge to an optimum solution. The weight vector is updated at each iteration by

$$\mathbf{w}(\kappa + 1) = \begin{cases} \left[\mathbf{I} - \gamma\mathbb{C}(\kappa)\right]\mathbf{w}(\kappa) + \gamma\mu\mathbf{A}_{n1}^{*} & \text{Howells-Applebaum} \\ \left[\mathbf{I} - \gamma\mathbb{C}(\kappa)\right]\mathbf{w}(\kappa) + \gamma\mathbf{s}^{\mathrm{T}} & \text{LMS} \end{cases} \tag{8.21}$$

where

$\mathbf{s}^{\mathrm{T}}$ = received signal
$\mathbf{I}$ = identity matrix
γ = step size
μ = constant

These algorithms are identical when the received signal is a constant times the main beam steering direction.

$$\mathbf{s}^{\mathrm{T}} = \mu\mathbf{A}_{n1}^{*} \tag{8.22}$$

See Ref. [6] for the derivations of (8.21). The Howells–Applebaum equation works in systems that know the desired signal direction in advance, for example, radar. It does not need to know the desired signal waveform. By contrast, the LMS equation does not need to know the desired signal direction but does need to know the desired signal waveform, for example, communications system.

The LMS convergence speed is proportional to the step size and is stable when the step size is bounded by [7]

$$0 \le \mu \le \frac{2}{p_{max}}. \tag{8.23}$$

where p_{max} is the maximum eigenvalue of the correlation matrix. In practice, the condition

$$0 \le \mu \le \frac{2}{p_{total}} \tag{8.24}$$

is used instead, since the total power entering the array (p_{total}) is easily computed, while the condition of Eq. (8.15) requires the computation of a covariance matrix and its eigenvalues. A small step size has slow convergence, while a large step size causes the algorithm to overshoot the optimum weights. Convergence is slow when the ratio of the maximum to minimum eigenvalues is large. Thus, the LMS algorithm converges quickly when the interference powers are similar and slow when they are not.

DMI algorithms are faster than the LMS algorithm and are not sensitive to the eigenvalue spread [6]. When the desired signal is absent, then the use of the beam-steering vector estimate yields the most desirable transient response characteristics. When the desired signal is strong; however, the beam-steering vector yields transient-response characteristics that are superior to the use of a priori information represented by the beam-steering vector alone.

8.4 ALGORITHMS FOR NONDIGITAL BEAMFORMERS

The previous sections described adaptive arrays that make use of a DBF. In those cases, all the beamforming is done in software. There are other approaches that are much less mathematically elegant but work with only amplitude and/or phase control at the elements. They have the advantage of operating in existing array architectures that do not use DBF. The cost savings is tremendous, and the calibration requirements are much less strenuous. In addition, these approaches are much more practical for arrays with many elements.

A non-DBF approach to adaptive nulling minimizes the total array output power. This notion seems ludicrous at first, because minimizing the total output power also minimizes the desired signal as well as the interference. Assuming that the desired signal enters the main beam and the interference enters the sidelobes, then it is possible to limit the adaptive weights to an extent that does little damage to the desired signal while eliminating the interference [11]. Non-DBF nulling creates nulls using only the hardware available in traditional RF analog beamforming system as shown

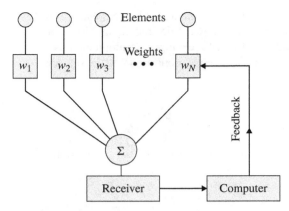

FIGURE 8.2   Non-DBF adaptive array.

in Figure 8.2. [12], [13]. All phased arrays have phase shifters. Some have attenuators as well. Large reductions in the amplitude weights or large phase shifts are required in order to reduce the main beam. Consequently, if only small amplitude and phase perturbations are allowed, then the main beam remains unscathed. Only small weight perturbations are needed to place nulls in the sidelobes, because the sidelobe levels are much lower than the main beam. A low sidelobe amplitude taper further reduces the perturbations to the weights needed to place the nulls [14].

Phase-only nulling was first proposed as a beam-space algorithm that used small phase shifts in a low sidelobe array to move nulls in the directions of interferers [15]. It has the advantage of simple implementation, because the only hardware needed is beam steering phase shifters. Small phase shifts produce a null on one side of the main beam, while simultaneously producing a higher sidelobe in the symmetric position on the other side of main beam. Allowing larger phase shifts [14] or adding amplitude control overcomes this problem.

Another approach takes the gradient with respect to the weights and uses the steepest descent algorithm to find the minimum output power [16]. As long as the weight changes are small and the sidelobes low, little main beam distortion occurs while placing the nulls. This approach has been implemented experimentally but is slow and can get trapped in a local minimum.

8.4.1   Partial Adaptive Nulling

Partial adaptive nulling uses a subset of elements to perform sidelobe nulling for the whole array [17], [18]. This approach reduces the number of adaptive weights which saves costs and increases convergence time. There must be enough adaptive elements to have a combined gain that is greater than or equal to the gain of the highest side-lobe [11]. If the first sidelobe is about 13 dB below the main beam, then more than $10^{(-13/20)} \times 32 \times 32 = 230$ elements are needed to place a null in the highest sidelobe.

A genetic algorithm (GA) was used to minimize the total output power of the array [19]. Results were averaged over 20 runs using 100–300 adaptive elements in the array to minimize the output power when a 0 dB signal at $(\theta,\phi) = (0°,0°)$ and 30 dB signal at $(\theta,\phi) = (4.4°,0°)$ are incident on the array.

Figure 8.3 shows a plot of the signal power received in the main beam and first sidelobe as a function of number of adaptive elements. The cross-over point when the desired receive signal exceeds the interference signal is at about 220 adaptive elements, which is close to the 230 elements previously estimated. Figure 8.4 is the SINR as a function of number of adaptive elements. Figure 8.5 is the adapted antenna pattern superimposed on the quiescent pattern at $\phi = 0°$ when 300 elements were adaptive. Figure 8.6 is a plot of the algorithm convergence. Each generation consists of no more than eight power measurements, so convergence is very fast.

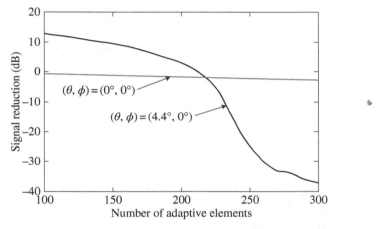

FIGURE 8.3   Signal reduction in the main beam and first sidelobe as a function of number of adaptive elements.

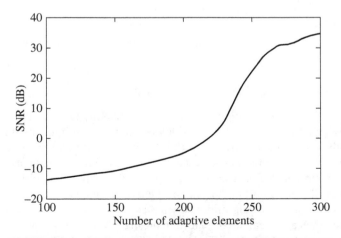

FIGURE 8.4   SINR as a function of number of adaptive elements.

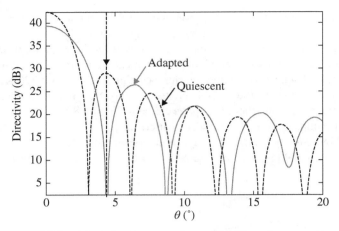

FIGURE 8.5 Adapted pattern superimposed on the quiescent pattern.

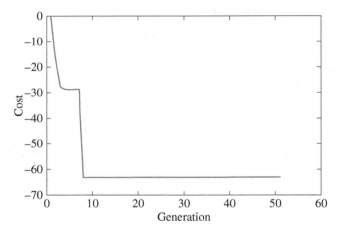

FIGURE 8.6 Algorithm improvement as a function of time.

8.4.2 Adaptive Nulling with Weight Constraints

Constraining the adaptive weights to small perturbations protects the main beam, while allowing nulls to be place in the sidelobes. A practical implementation of this strategy only allows the adaptive algorithm to have access to a few least significant bits of a digital phase shifter and attenuator [20]. Large bits are quite capable of disrupting the main beam. For instance, giving half the elements a 180-degree phase shift (bit 1) will place a null in the peak of the main beam. On the other hand, if the output power minimization algorithm only has access to the 22.5° phase bit of a 4-bit phase shifter, then a null cannot be placed in the main beam. This single LSB reduces the sidelobe as shown in Figure 8.7, but is not potent enough to place a null. The 45°

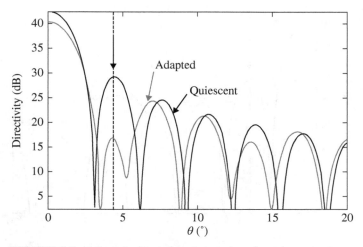

FIGURE 8.7   Adapted pattern superimposed on the quiescent pattern.

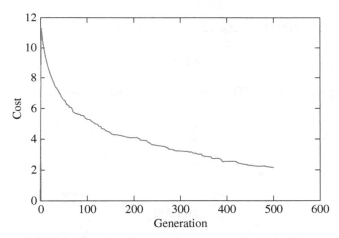

FIGURE 8.8   Algorithm improvement as a function of time.

bit would also have to be used to place a null. Using a second bit would increase distortion to the pattern, though. Figure 8.8 shows the convergence of the genetic algorithm for adapting.

8.4.3   Adaptive Nulling with Cancellation Beams

For a planar array, the weights needed to place M nulls in the pattern are given by [21]

$$w_n = 1 - \sum_{m=1}^{M} \gamma_m e^{-jk(x_n u_m + y_n v_m)} \tag{8.25}$$

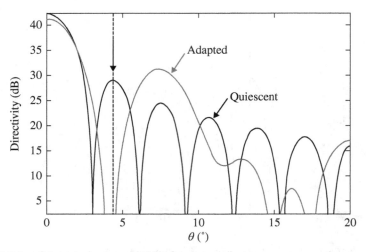

FIGURE 8.9 Adapted pattern superimposed on the quiescent pattern.

If the adaptive weights have adjustable phase only, then

$$w_n = e^{j\delta_n} \approx 1 + j\delta_n \tag{8.26}$$

Substituting (8.26) into (8.25) and solving for the phase yields

$$\delta_n = j\sum_{m=1}^{M} \gamma_m e^{-jk(x_n u_m + y_n v_m)} \tag{8.27}$$

The phase must be real, so

$$\delta_n = \sum_{m=1}^{M} \gamma_m \sin\left[k(x_n u_m + y_n v_m)\right] \tag{8.28}$$

The phase can be adaptively adjusted to place a null by varying the variable γ_m. Figure 8.9 shows the adapted pattern, while Figure 8.10 shows the convergence when a genetic algorithm is used to minimize the total output power. This technique produces a lot of sidelobe distortion. Using a subset of the elements or a low sidelobe amplitude taper would reduce those distortions.

8.5   RECONFIGURABLE ARRAYS

Reconfigurable arrays alter the current flow on their radiating elements using mechanically movable parts, phase shifters, attenuators, diodes, tunable materials, or active materials in order to change the array's performance characteristics [22].

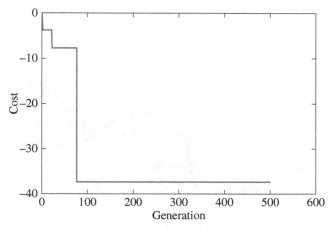

FIGURE 8.10   Algorithm improvement as a function of time.

The performance characteristics to be adapted may include the antenna pattern, polarization, or frequency/bandwidth in some desirable fashion. Some examples include placing a null in an antenna pattern, switching from right-hand circularly polarized to left-hand circularly polarized operation, or moving the carrier frequency from 1 to 3 GHz. A reconfigurable array may either have reconfigurable elements or may change the amplitude and phase, the number, the positions, or the polarization of the elements. Spherical arrays, radio astronomy arrays, and formation flying arrays are three types of reconfigurable arrays.

Beam steering with a spherical array is done by reconfiguring the active elements on the array. Spherical arrays ideally scan the hemisphere by activating elements in a small area of the sphere whose normal points approximately in the desired direction. Electronic beam steering further refines the pointing direction. This complicated beam steering on a spherical array is rewarded with the following [23]:

1. Lower polarization losses
2. Lower mismatch losses
3. No beam shift versus frequency (implies wider bandwidth)
4. No gain loss with beam steering

A planar array maximum scan angle is less than 60 degrees, while a spherical array's is not limited. On the down side, spherical arrays are expensive to build, because manufacturing the curved surfaces and associated feed networks is challenging.

Approximating the sphere by a polyhedron (a solid figure with four or more faces) allows the use of standard planar array technology for panels on the facets of the polyhedron. The resulting geodesic dome is mechanically stable and supports itself without needing internal columns or interior load-bearing walls.

Figure 8.11 shows one version of a geodesic dome array. Each triangle on the dome is a subarray that is either active or inactive [23]. Groups of active subarrays form an

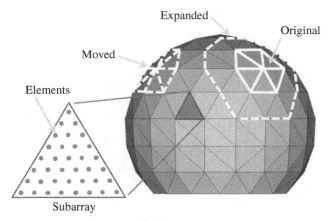

FIGURE 8.11 A phased array built on a geodesic dome. The outlined triangles represent regions of activated subarrays. The solid region is the original array configuration. The dashed outlines are either the expanded or translated version of the original configuration.

aperture large enough to have the desired gain. Adding more subarrays to expand the original array (Fig. 8.11) increases the gain and narrows the beamwidth. Moving the active aperture to another part of the dome points the main beam in a different direction. Phase shifters at the elements perform the beam pointing direction.

A geodesic dome may also be made from hexagons and pentagons having the same side lengths. A section of a 10 m diameter geodesic dome array prototype was recently built for tracking and communicating with satellites (Fig. 8.12). The section had one pentagonal and five hexagonal panels. Each panel comprised hexagonal sub-arrays with 37 circularly polarized transmit/receive elements in a triangular lattice [25]. A pentagon panel had 10 subarrays (370 elements), while a hexagon panel had 21 subarrays (777 elements). The array operated near 2 GHz.

Some very large reconfigurable radio telescopes move large reflector antenna elements in order to configure an ideal layout for observations. For instance, the Atacama Large Millimeter/submillimeter Array (ALMA) is a radio telescope with 66 12 and 7 m parabolic dishes in Chile's Andes Mountains (Fig. 8.13) [27]. The telescope operates at frequencies from 31.25 to 950 GHz. Specially designed vehicles move the ALMA antennas between flat concrete slabs to create array configurations that extend from 250 m to 15 km. A second example of physically moved elements in an array is the very large array (VLA) radio telescope in New Mexico made up of 27 dish antennas that are 25 m in diameter [28]. Elements are placed in a Y-shape and can be changed by moving the elements along a railroad track (Fig. 8.14). The array operates from 73 MHz to 50 GHz in four different aperture sizes of 36, 10, 3.6, or 1 km.

Reconfigurable arrays may also be made from antennas placed on formations of unmanned aerial vehicles, small satellites, or robots on the ground. One example of an array of elements made from a formation of small satellites is the TechSat 21 project. It was a radar concept that consisted of a cluster of 20 small satellites (Fig. 8.15), each having a 4 m^2 array that operates at X band [29]. Each satellite

FIGURE 8.12   Section of a 10 m diameter spherical array [24]. © 2010 Wiley.

FIGURE 8.13   Atacama large millimeter/submillimeter array. Reprinted by permission of Ref. [26]; © 2013 IEEE.

FIGURE 8.14   VLA antennas on railroad tracks (NRAO/AUI/NSF).

FIGURE 8.15   TechSat21 concept. ("TechSat21." Licensed under Public domain via Wikimedia Commons—http://commons.wikimedia.org/wiki/File:TechSat21.jpg#mediaviewer/File:TechSat21.jpg)

transmits orthogonal signals at different frequencies but receive the return signals from all the satellites. In order for the sparse aperture to coherently form beams, the antenna phase centers (and, hence, satellite positions) must be known to within a fraction of a wavelength which is difficult to achieve in practice.

8.6 RECONFIGURABLE ELEMENTS

Another approach to reconfiguring an array is to reconfigure the elements. Switches in the elements connect and disconnect parts of the antenna to get desirable operating characteristics. As an example, switches can be used to reconfigure microstrip patch antennas in several different ways. For instance, Figure 8.16 shows a two-dimensional array of metal patches on a substrate [30]. The sides of adjacent patches are connected with RF switches in order to configure a patch antenna into a desired configuration. Figure 8.17 is an example of a patch with slots that can be shorted in order to switch between polarizations [31].

MEMS switches in reconfigurable antennas started in the late 1990s [32]. Figure 8.18 are examples of two UWB ultra wideband monopoles with a reconfigurable band notch in the wireless LAN frequency range (5.150–5.825 GHz) are shown in [33]. A coplanar waveguide line feeds the $\lambda/2$ U-shaped slot. There is a frequency notch when the MEMS switch is open but not when the MEMS switch is closed. When the switch is open at 5.8 GHz, the currents in the inner and outer side of the slot flow in opposite directions and cancel each other. When the MEMS switch is closed, the slot is short at its center point, so the total slot length is cut in half. The slot no longer supports the resonating currents and radiation occurs as if the slot is not there. A second reconfigurable antenna has two inverted L-shaped open stubs connected by MEMS switches to the elliptical patch, which operates over 3.1–10.6 GHz). Shorting

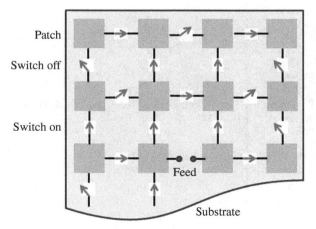

FIGURE 8.16 Reconfigurable microstrip patches. Reprinted by permission of Ref. [15]; © 2013 IEEE.

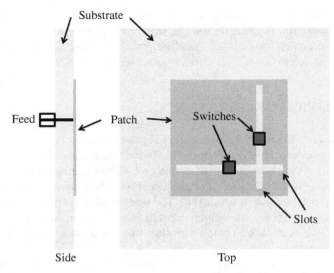

FIGURE 8.17   Reconfigurable slotted patch antenna. Reprinted by permission of Ref. [15]; © 2013 IEEE.

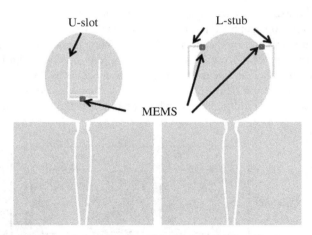

FIGURE 8.18   Diagram of the reconfigurable U-slot and L-stub antennas. Reprinted by permission of Ref. [15]; © 2013 IEEE.

the stubs to the patch creates a rejection band. At the resonance frequency, the direction of the current on the stub flows in the opposite direction of the current on the nearby antenna edge and cancels the radiated field. When the stubs are not shorted, the antenna has no rejection band. The MEMS switches are actuated through the RF signal path, without any dc bias lines.

Reconfigurable antennas can also change the antenna polarization. For instance, placing MEMS switches in the feeds of microstrip antennas provides the ability to

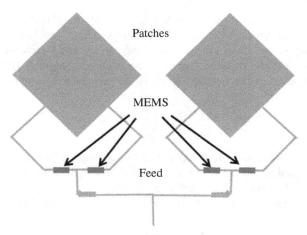

FIGURE 8.19 Reconfigurable antenna that change polarization using MEMS. Reprinted by permission of Ref. [34]; © 2013 IEEE.

switch from one linear polarization to the orthogonal linear polarization or to circular polarization as shown in Figure 8.13 [34] (Fig. 8.19).

Varactor diodes have been used to tune the operating frequencies of an antenna. As an example, varactors were placed at the radiating edges of a microstrip patch antenna in order to increase its bandwidth to 30% [35]. Strategically placing varactors along a slot antenna produces dual bands with controllable first and second resonant frequencies [36]. A reconfigurable partially reflecting surface (PRS) antenna has been built that has a measured realized gain from 10 to 16.4 dBi over the tuning range of 5.2 to 5.95 GHz, as the bias voltage was tuned from 6.49 to 18.5 V [37]. A slotted monopole antenna, modified with a varactor, can be tuned between 1.6 and 2.3 GHz without a significant change in the antenna efficiency [38]. Radiation pattern control has also been demonstrated using varactors [39].

8.7 TIME-MODULATED ARRAYS

Amplitude and phase errors prevent small arrays from having very low sidelobes. In 1963, a time-modulated array was introduced in which synchronized switches opened and closed at the array elements in order to connect and disconnect them from the feed network, in such a way that the average antenna pattern for a small array had low sidelobes [40]. The time-modulated, time-averaged low sidelobe pattern gain is less than the gain of an equivalent static (no switching) amplitude tapered array [41]. As an example, an array with a time modulated 40 dB Chebyshev sidelobe taper has 3.53 dB less gain than a 40 dB Chebyshev sidelobe static amplitude tapered array. Basically, a 1 dB loss in directivity means that 20% more elements are needed to get the same gain from the array.

When the switch is closed, the element connects to the feed network ($w_n = 1.0$). If the switch is open ($w_n = 0.0$), the element contributes no signal to the output. The switch closed time, τ_n, determines the average element weights. Each switch rapidly opens and closes in synchronization to an external periodic signal having a period of T_s. An element whose switch is always closed has an average weight of 1.0, while an element whose switch is always open has an average weight of 0.0. The average amplitude weight at element n is the fraction of the period that a switch is closed, so that

$$\bar{w}_n = \frac{\tau_n}{T_s} = a_n \qquad (8.29)$$

where a_n is the desired amplitude taper. The switches are assumed to not impact the phase of the signal. Once the switching period is known, then the switch closed time is given by $\tau_n = T_s a_n$ for each element (Fig. 8.20).

The new time-varying weights lead to the following time-dependent array factor [40]:

$$AF = \sum_{p=-\infty}^{\infty} e^{j[(\omega + p\omega_s)t]} \sum_{n=1}^{N} w_{pn} e^{jk(n-1)du} \qquad (8.30)$$

where

$$w_{pn} = \frac{1}{T_s} \int_0^{T_s} w_n(t) e^{-jp\omega_s t} dt$$

$$w_n(t) = w_{N+1-n}(t) = \begin{cases} 1 & 0 \le t \le \tau_n \\ 0 & \tau_n < t \le 1 \end{cases}$$

$$\omega_s = \frac{2\pi}{T_s}$$

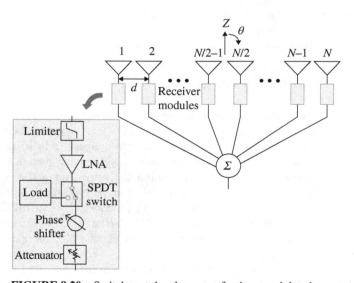

FIGURE 8.20   Switches at the elements of a time-modulated array.

Symmetric weights about the array center are assumed equal. This equation implies that the array factor not only exists at the fundamental frequency, ω_s, but array factors also exist at all the harmonics of the switching frequency, $p\omega_s$.

In the time domain, when signals have a bandwidth and the element weights are a function of time, then the total signal received by an array with a nondispersive feed network is found from

$$F = \sum_{n=1}^{N} w_n(t) \sum_{m=1}^{M} s_m \left[t - (n-1)\frac{d}{c}\sin\theta_m \right] \qquad (8.31)$$

where s_m = strength of signal m incident at θ_m, M = number of signals incident on the array, and c = speed of light

In this formulation, when $\theta_m = 90°$, then element N receives the signal first and element 1 receives it last. This equation is a better formulation for investigating the time behavior of the array than is (8.30), because modulated signals can be modeled, beamsteering effects can be investigated, and interference can be included.

Time modulated arrays assume that an "average" array pattern receives time-varying signals; whereas in reality, the instantaneous uniform array factors actually receive the signal [42], [43]. This uniform array factor is a function of the number of element switches that are closed at that instant. As an example, consider an eight-element linear array with $d = \lambda/2$. Amplitude weights and switch times are assumed to be symmetric about the center of the array. Table 8.1 shows the amplitude weights and corresponding switching times for 30 and 40 dB Chebyshev average patterns when the carrier frequency is 300 MHz and $f_s = 15$ MHz ($T_s = 66.7$ ns). This switching time is consistent with switches that are available today. One period of the array factor of the 30 dB average Chebyshev array is shown in Figure 8.21. Note that there are four distinct uniform array factors corresponding to eight-, six-, four-, and two-element arrays that exist for 17.5, 17.1, 19.5, and 12.5 ns, respectively. The switching times change for the 40 dB Chebyshev average taper. Now, the eight-, six-, four-, and two-element uniform arrays exist for 9.7, 18.1, 22.7, and 16.1 ns, respectively. Figure 8.22 shows a time-angle plot of the array factor over one switching period. The four different uniform array factors are identical in Figure 8.21 and Figure 8.22, but the length of time that they exist differs. In either case, when time is frozen, only

TABLE 8.1 Element Weights and Switching Times for 30 and 40 dB Chebyshev Average Array Factors

SLL		Element			
		1	2	3	4
−30 dB	w_n	0.262	0.519	0.812	1.000
	τ_n (ns)	17.47	34.60	54.14	66.67
−40 dB	w_n	0.146	0.418	0.759	1.000
	τ_n (ns)	9.73	27.87	50.60	66.67

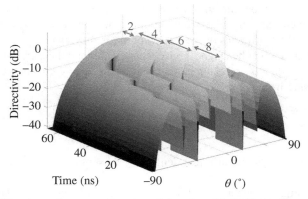

FIGURE 8.21   Array factor as a function of time for a 30 dB Chebyshev average taper.

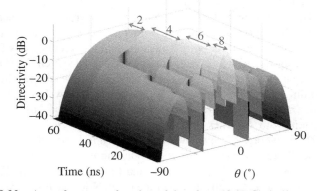

FIGURE 8.22   Array factor as a function of time for a 40 dB Chebyshev average taper.

a uniform array factor exists and not a low sidelobe array factor. The "low sidelobes" result from averaging these instantaneous array patterns.

The sidelobes, nulls, and main beam of the array factor change as the number of elements with closed switches change. Figure 8.23 is a plot of the main beam directivity as a function of time. This plot shows that the directivity varies from 9 to 7.8 to 6 to 3 dB corresponding to eight, six, four, and two elements. Assuming that a desired signal enters the main beam, then the amplitude of the desired signal is modulated by the time-varying main beam. As a result, the total signal received by the array is also modulated.

Figure 8.24 compares the time-modulated array factor for the eight-element array with uniform and 30 dB Chebyshev array factors over two switch periods. The top plot of this figure shows the amplitude weights of the time-modulated array as a function of time. The array starts as an eight-element uniform array then changes to six-, four-, and finally two-element uniform arrays before repeating. The

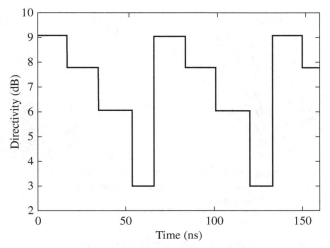

FIGURE 8.23   The directivity of the array factor as a function of time for a 30 dB Chebyshev average taper.

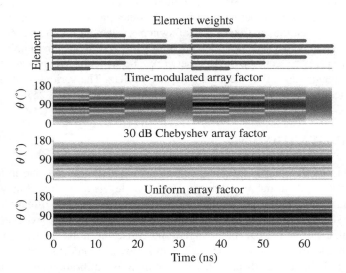

FIGURE 8.24   The top plots are the weights and array factor (in dB) for the time-modulated array for a 30 dB Chebyshev average. The bottom two plots are static 30 dB Chebyshev and uniform array factors.

time-modulated array factor always has higher sidelobes than the Chebyshev pattern (darker means higher level than lighter shade). In addition it also has a lower gain and wider main beam for a significant amount of time. The Chebyshev pattern has a directivity of 7.85 dB, while the time-modulate array has a directivity of 9, 7.78, 6, and 3 dB over T_s.

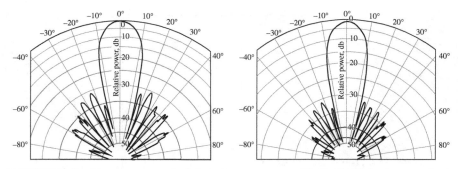

FIGURE 8.25   The pattern on the left is from an array with 30 dB Chebyshev amplitude weights. The one on the right is a time-modulated 30 dB Chebyshev taper. Reprinted by permission of Ref. [40]; © 1963 IEEE.

Time modulation was experimentally demonstrated with an eight-element slot array with three diode switches shared by symmetric elements [40]. The two center elements were always connected to the feed network, while the other switches were on for a time that was proportional to the desired amplitude weight. A monotonically decreasing time averaged amplitude taper resulted when the outer pair (element numbers 1 and 8) switches open, then the next outer pair of switches (element numbers 2 and 7) open, and finally the next two switches (element numbers 3 and 6) open until only the center pair (4 and 5) are connected to the feed network. The switching frequency was 10 kHz, while the antenna operating frequency was 9.375 GHz. The various switching times were selected to create a time-averaged low sidelobe pattern (Fig. 8.25).

8.8   ADAPTIVE THINNING

Chapter 3 introduced array thinning. Making some elements inactive in a uniform aperture cause nulls to move and average sidelobe levels to decline. Many different random thinnings result in the same statistical sidelobe level and directivity, but the individual sidelobe and nulls are different. If all the elements have switches, then the elements that are active can be changed and cause the nulls to move adaptively. Elements in the array must have active T/R modules with on/off switches in order to be adaptive. Switches in T/R modules are common for calibration and are much less expensive to implement than the RF ADCs needed for DBF, resulting in a low-cost adaptive array. Adaptive thinning has the advantages of being high speed, not changing the array directivity, requiring switches instead of ADCs, and needing no special calibration. Its major disadvantage is the inability to null the main beam and possibly first sidelobe. Adaptive thinning for nulling works best for large arrays, much as low sidelobe thinning works best for large arrays.

Dynamic thinning is a nonadaptive approach that switches between preselected thinning configurations in order to place nulls in the directions of interference [44], [45]. The thinning configurations are selected to produce broad nulls that cover specified angular regions. Randomly thinned arrays of the same size have the same or nearly the same average sidelobe level [46]. By forcing the number of elements turned on to be the same, then the array directivity does not change when the nulls and sidelobes do. They then presented a method for dynamically thinning a linear array in order place low sidelobes and nulls in desired directions. In [47], the thinning was adapted to raise or lower the sidelobe levels of a planar array. If there is no interference, then high sidelobe levels mean high directivity, so the array is more sensitive to the desired signal. When interference is present, then the sidelobe level is lowered via additional thinning to mitigate the interference at the cost of a small decrease in directivity. In [48] RF switches connect or disconnect array elements in order to generate nulls in the directions of the interfering signals. A genetic algorithm maximizes the signal-to-interference plus noise ratio (SINR). Two variants were developed: one does not constrain the number of active elements, while the other forces the solution to satisfy a fixed-directivity criterion.

A thinned configuration that mimics a desired low sidelobe taper is not unique. Since the active and inactive elements are randomly selected, changing the seed of the random number generator results in a different pattern of active and inactive elements. Even though the RMS sidelobe levels and directivity are the same for the different thinnings, the sidelobes and nulls change. New random thinnings are tried until one is found that has nulls in the directions of the interfering signals and minimizes the total output power [49], [50].

The key to the adaptive thinning approach is the T/R module in Figure 8.26. This module has a single-pole triple throw (SP3T) absorptive switch that selects either the transmit path, the receive path, or a matched load. Connecting to the load disconnects the element from the feed network and turns the element off.

Consider the case where two interfering sources are incident on a 32x32 element planar array at $\theta = 14°$ and $41.5°$ when $\phi = 0°$. The SINR is calculated for each of 100 random thinnings that imitate an aperture with a 25 dB $\bar{n} = 5$ Taylor taper (Fig. 8.27). Three different scenarios are shown in the plots. The first has both interferers at 20 dB above the desired signal entering the main beam. A line is drawn at

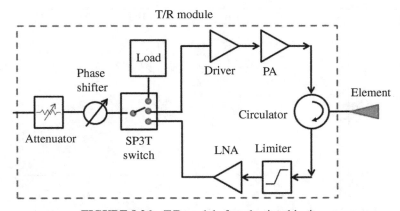

FIGURE 8.26 T/R module for adaptive thinning.

0 dB. Thinnings below that line are considered unacceptable (0 dB is an arbitrary choice). The 20 dB case has 90% of the thinnings above the 0 dB mark. The second case increases both interferers to 25 dB. Now, about half the thinnings are above 0 dB SINR, so 50% of the time, a thinning that mimics the desired amplitude taper results in an SINR>0 dB. The third case has 2 interferers that are 30 dB above the desired signal. Only 2 random thinnings place nulls of sufficient depth to get an SINR>0 dB. Searching for the best random thinning to reduce interference takes longer as the interference sources increase their power.

The best and worst cases for the thinnings in Figure 8.27 are shown in Figures 8.28 and 8.29. The worst case has sidelobes in the directions of the interferers, while the

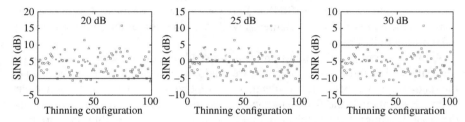

FIGURE 8.27 SINR for 100 random seeds that generate a thinning for a 25 dB Taylor taper.

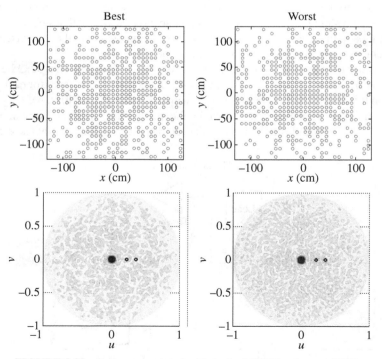

FIGURE 8.28 Pattern associated with the best SINR and worst SINR.

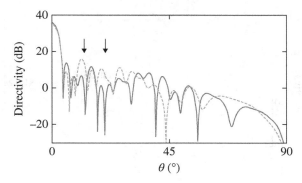

FIGURE 8.29 Pattern cuts at $\phi = 0°$ for the best and worst SINR cases.

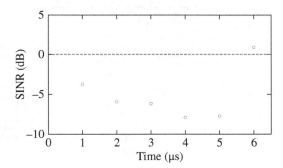

FIGURE 8.30 SINR as a function of switching time for a 25 dB Taylor taper when two interference sources are 30 dB stronger than the desired signal.

best case has nulls in those directions. The arrangement of the active elements is quite different, yet the number of active elements (and directivity) is approximately equal. Null locations change dramatically, while the sidelobe levels remain about the same as shown by the $u–v$ plots.

In order to show the previous example as an adaptive process, the random number seed is changed, and thinnings are generated until the $SINR>0$ dB. The switching time in the T/R modules is assumed to be $1\mu s$. The convergence plot in Figure 8.30 shows that the SINR goes above zero after six random changes to the thinning, so it is converged. If the interference sources were 35 dB stronger than the desired signal, then the SINR goes above zero after 41 random thinnings as shown in Figure 8.31. If the interference is too strong, then no random thinning will raise the SINR above zero and the system is jammed. A computer model that includes mutual coupling and array errors would produce to a plot similar to Figure 8.30, but the circles would not be in the same positions. Pattern cuts are compared in Figure 8.32. The worst SINR plot has a sidelobe in the directions of the interfering signals (arrows), while the best SINR plot has a null or reduced sidelobe in those directions.

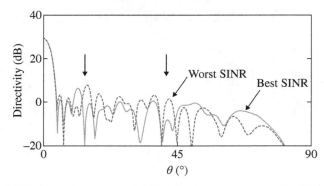

FIGURE 8.31   Pattern cuts at $\phi = 0°$ for the best and worst SINR cases.

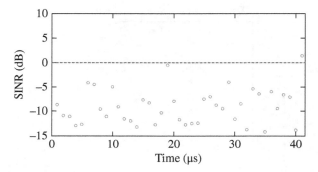

FIGURE 8.32   SINR as a function of switching time for a 25 dB Taylor taper when two inter-ference sources are 35 dB stronger than the desired signal.

For the second example, consider a concentric ring array with eight rings. The rings and elements in the rings are separated approximately 0.5λ. Figure 8.33 shows two random thinnings for a 30 dB $\bar{n} = 5$ Taylor taper. Only the outer two rings have inactive elements. If thinning 1 does not eliminate the interference, then the random thinning is changed until the nulls are placed in the far-field pattern. Figure 8.33 shows the starting and ending thinning configurations. Both thinnings result in approximately the same peak sidelobes as shown by the pattern cuts in Figure 8.34. If interference is incident on the sidelobes at $\theta = 41°$ and $70°$ when $\phi = 0°$, then thinning 2 would place nulls in the directions of the interfering signals. The interference at $\theta = 41°$ is reduced by over 20 dB and the interference at $\theta = 70°$ is reduced by 14 dB.

The following are the advantages of adaptive thinning over other adaptive nulling techniques:

- No DBF is required.
- It works best for large arrays, while other adaptive procedures address small arrays.
- The adaptive algorithm uses simple random guessing.
- The algorithm is fast—limited by switch speed.

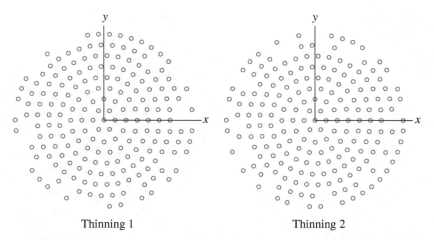

Thinning 1   Thinning 2

FIGURE 8.33   Two random low sidelobe thinnings for the concentric ring array.

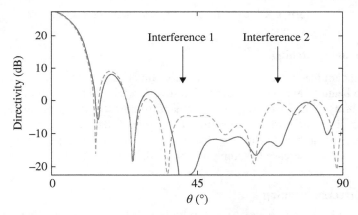

FIGURE 8.34   The red dashed line is a pattern cut of the array with thinning 1, while the blue solid line is the pattern cut from thinning 2.

- Hardware requirements are minimal.
- It maintains main beam gain
- Sidelobe distortion is minimal

These advantages are important for implementing adaptive nulling on very large arrays.

8.9   OTHER ADAPTIVE ARRAY ALTERNATIVES

This section presents some other important adaptive array concepts. Beam switching and MIMO enhance reception of a signal. Direction finding or angle of arrival estimations locates signals in space, while a retrodirective array functions as a transponder.

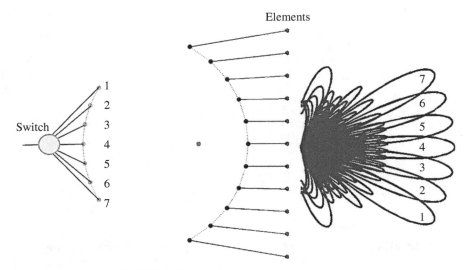

FIGURE 8.35   Rotman lens with multiple beams.

8.9.1   Beam Switching

Multiple beam feeds like the Rotman lens or Butler matrix allow an array to switch from one beam to another in order to improve reception. Figure 8.35 shows an 11-element array with seven orthogonal beams formed by the Rotman lens. Beam switching picks the port corresponding to the beam that best receives the desired signal. In general, beam switching does not mitigate interference entering the side-lobes, but it does increase the amplitude of the desired signal.

8.9.2   Direction Finding

Arrays are ideal platforms for estimating the angle of arrival of signals. The main beam is one way to detect the signals—a strong output indicates the presence of a signal. This approach does give very accurate angle information, because main beams are wide in angle unless the array is very large. Nulls, on the other hand, are narrow and can be used to very accurately locate a signal. When the output power is minimized, then the signals are in nulls.

A circular or Wullenweber array uses a commutating feed to steer a beam 360° around in azimuth to locate signals [51]. The angle at which the received signal is largest gives the azimuth of the signal. Multiple signals that are closely spaced may be difficult to resolve. The plot of signal strength versus angle is a type of direction finding spectrum called a periodogram [52]. Direction finding with a main beam is also possible using the multibeam Rotman lens and interpolating between beams using a neural network [53].

An Adcock array has four elements on the corners of a square. Sum and difference beams are generated by combining the two diagonal element pairs as shown in

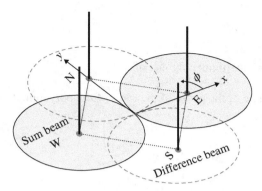

FIGURE 8.36   Diagram of an Adcock array.

Figure 8.36 [54]. One diagonal element pair has a pattern that is orthogonal to the pattern of the other element pair. The signal direction is calculated from the ratio of the signal received by the difference to the signal received by the sum pattern. Nulls are in a fixed position, so no real adaptation is required.

In general, direction finding arrays move nulls in the directions of all signals by adjusting the weighting at each element then calculating the location of the nulls from the resultant weights. The signal processing algorithms are similar to adaptive nulling algorithms, except that all signals are interference and require nulls in those directions. Superresolution algorithms use nulls to locate signals, so their direction finding spectra have narrow peaks in the directions of the signals.

If two signals lie less than a beamwidth apart, then they are difficult to resolve in a periodogram. DBF direction finding arrays form the correlation matrix and then employ superresolution techniques to resolve closely spaced signals. One example is the Capon superresolution spectrum [55] given by

$$P(\theta) = \frac{1}{A^\dagger(\theta)\, \mathbb{C}^{-1}\, A(\theta)} \tag{8.32}$$

Another alternative is the *MU*ltiple *SI*gnal *C*lassification (MUSIC) [56] spectrum with the spectrum

$$P(\theta) = \frac{A^\dagger(\theta)\, A(\theta)}{A^\dagger(\theta)\, V_\lambda\, V_\lambda^\dagger\, A(\theta)} \tag{8.33}$$

where V_λ is a matrix whose columns contain the eigenvectors of the noise subspace. The eigenvectors of the noise subspace correspond to the $N - N_s$ smallest eigenvalues of the correlation matrix. The denominator of (8.33) can be written as a polynomial

$$A^\dagger(\theta)\, V_\lambda\, V_\lambda^\dagger\, A(\theta) = \sum_{n=M+1}^{M-1} c_n z^n \tag{8.34}$$

where

$$z = e^{jknd\sin\theta}$$

$$c_n = \sum_{r-c=n} V_\lambda V_\lambda^\dagger = \text{sum of } n\text{th diagonal of } V_\lambda V_\lambda^\dagger$$

Solving for the angle of the phase of the roots of the polynomial in (8.34) yields the signal locations at

$$\theta_m = \sin^{-1}\left(\frac{\arg(z_m)}{kd}\right) \tag{8.35}$$

The maximum entropy method (MEM) spectrum is given by [57]

$$P(\theta) = \frac{1}{A^\dagger(\theta)\mathbb{C}^{-1}[:,n]\mathbb{C}^{-1\dagger}[:,n]A(\theta)} \tag{8.36}$$

where n is the nth column of the inverse correlation matrix.

An eight-element array of isotropic point sources spaced $\lambda/2$ apart and lying along the x-axis serves as the test bed for four different spectra. Sources are incident on the array at $-60°$, $0°$, and $10°$ with relative powers of 0, 4, and 12 dB, respectively. The periodogram has broad peaks and cannot distinguish the sources at $0°$ and $10°$. Capon, MUSIC, and MEM spectra have very sharp spikes in the directions of the sources and can distinguish closely spaced sources (Fig. 8.37). The superresolution techniques are very sensitive to errors, multipath, and signal strength and only work with a DBF. Even though the periodogram suffers from angular resolution problems, it is very robust and does not require a DBF.

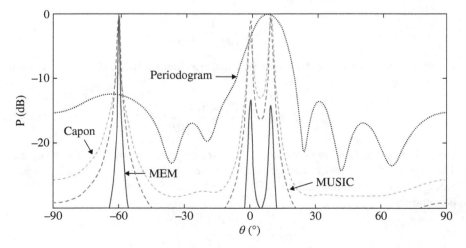

FIGURE 8.37 Plot of the direction finding spectra for an eight-element array.

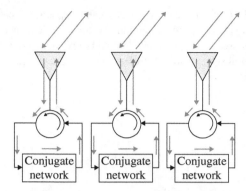

FIGURE 8.38   Diagram of a retrodirective array.

8.9.3   Retrodirective Array

A retrodirective array retransmits a received signal in a desired direction. Figure 8.38 shows a retrodirective array that receives a signal, takes the complex conjugate, amplifies, and then retransmits it [58]. Retrodirective arrays have applications in radio frequency identification (RFID) [59], wireless communications, and RF transponders where a signal must be returned to the source. The signal may be amplified and the frequency or polarization may be slightly altered to distinguish it from the original signal. It is possible to have a two-dimensional retrodirective array that responds to both azimuth and elevation angles [60], [61].

8.9.4   MIMO

Almost all radar and communication systems use a single input–single output (SISO) architecture. A signal is sent from a transmit antenna to a receive antenna. The transmit and receive beams are optimized for line-of-sight (LOS) transmission and reception: maximize the main beam to enhance the desired signal and minimize sidelobes to reject interference. Multipath and obstacles are detrimental in SISO, since they cause fast and slow fading that decreases signal strength and increases noise.

   If the transmit and receive antennas are arrays, then each element in the transmit array sends a signal to each element in the receive array. The relationship between the signals sent by the transmit array ($\mathbf{s}_t$) and the signals received at the elements of the receive array ($\mathbf{s}_r$) are defined by

$$\mathbf{s}_r = \mathbf{s}_t \mathbf{H} + \mathbb{N} \tag{8.37}$$

where the channel matrix given by

$$\mathbf{H} = \begin{bmatrix} h_{11} & h_{12} & \cdots & h_{1N} \\ h_{21} & h_{22} & & \vdots \\ & & \ddots & \\ h_{M1} & \cdots & & h_{MN} \end{bmatrix} \tag{8.38}$$

Each element in $\mathbf{H}$ (h_{mn}) characterizes a channel propagation path between transmit element m and receive element n (channel transfer function). The received data is the transmitted data after it passes through the channel and picks up noise (N). In free space with no multipath (typical SISO scenario), the elements of $\mathbf{H}$ are the free space Green function.

$$\mathbf{H} = \begin{bmatrix} \dfrac{e^{-jkR_{11}}}{R_{11}} & \dfrac{e^{-jkR_{mn}}}{R_{mn}} & \dfrac{e^{-jkR_{mn}}}{R_{mn}} \\[2ex] \dfrac{e^{-jkR_{21}}}{R_{21}} & \dfrac{e^{-jkR_{mn}}}{R_{mn}} & \\[4ex] \dfrac{e^{-jkR_{mn}}}{R_{MN}} & & \dfrac{e^{-jkR_{mn}}}{R_{mn}} \end{bmatrix} \tag{8.39}$$

where R_{mn} is the distance between transmit element m and receive element n.

Theoretically, an estimate of the transmitted data, $\mathbf{s}_t$, can be found in a noise free environment by inverting the channel matrix and multiplying the received data.

$$\mathbf{s}_t = \mathbf{H}^{-1}\mathbf{s}_r \tag{8.40}$$

In a SISO system, however, (8.39) is ill-conditioned, so a good estimate of $\mathbf{s}_t$ is not possible. In any event, these equations assume that the transmit and receive arrays use DBF.

The channel matrix in (8.39) assumes a direct line of sight between all the element with no multipath components. A Rician (LOS plus multipath) or Rayleigh (no LOS but only multipath) channel has a channel matrix that is not ill-conditioned, because the multipath adds random signals to each element. Increasing the separation between the arrays as well as increasing the element spacing in the arrays creates more opportunity for multipath. Large element spacings in the base station are usually not a problem, but element spacing at the mobile unit is severely limited.

Multiple input–multiple output (MIMO) was developed for communications in a high multipath (Rayleigh fading) environment [62]. Figure 8.39 is a diagram of the

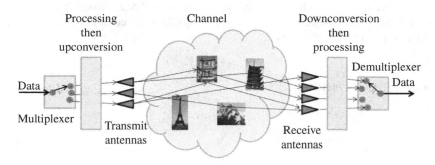

FIGURE 8.39   MIMO concept.

MIMO concept [63]. Transmit data is divided between all the transmit elements of the array. Thus, each element transmits a different data stream. The processing precodes the data, so that the receive elements can recognize which transmit element sent the data as well as correct errors in the bits. Spatial multiplexing increases the data transmission rate while using the same bandwidth and power as a SISO system. All of these signals arrive at each antenna in the receive system via different paths. The received signals are down-converted, filtered, sampled, and decoded to get the original message.

The theoretical capacity of a MIMO system is proportional to the number of transmit/receive antenna pairs in the MIMO system [64]. A MIMO system must have at least as many receivers as data streams transmitted. The number of these transmit streams should not be confused with the number of transmit antennas. If there are N_r transmit antennas and N_t receive antennas, then the capacity improvement is proportional to the smaller number, N_r or N_t. Spatial multiplexing increases data rates in a high multipath environment, due to low correlations between the channels. Highly correlated channels cause the spatial multiplexing performance to rapidly degrade.

Multipath, noise, fading, Doppler shift, coupling, and interference cause variations to the matrix elements that are difficult to predict. In order for MIMO to work, the transmit array sends a burst of unique training symbols from each element that allows the receiver at each receive element to get an accurate measure the propagation conditions between all combinations of elements in the transmit and receive arrays [65]. The measured data is used to form a channel matrix (assuming flat fading where all frequency components experience the same effects). When possible, this information is then fed back to the transmitter, so both the transmit and receive antennas have full channel information. Feeding the channel information back to the transmit antenna in real time is difficult. Channel capacity can still be significantly increased through a slow feedback of an average channel matrix or totally ignoring the feedback altogether.

Diversity Coding techniques are used when there is no channel knowledge at the transmitter. In diversity methods, a single stream (unlike multiple streams in spatial multiplexing) is transmitted, but the signal is coded using techniques called spacetime coding. The signal is emitted from each of the transmit antennas with full or near orthogonal coding. Diversity coding exploits the independent fading in the multiple antenna links to enhance signal diversity. Because there is no channel knowledge, there is no beamforming or array gain from diversity coding. Diversity coding can be combined with spatial multiplexing when some channel knowledge is available at the transmitter.

A MIMO radar transmits independent waveforms from N_t elements or subarrays [66]. The signal scatters from the target and returns to the same elements or subarrays or to different element or subarrays. Statistical or noncoherent MIMO exploits the random target radar cross section fluctuations with respect to angle that cause spatial decorrelation of the target returns and precluding coherent processing. Statistical MIMO applies to widely separated antennas, such as a bistatic radar. Coherent MIMO radar has antenna elements that are close together, so the target returns are correlated.

REFERENCES

[1] R. L. Haupt, "Adaptive arrays," *ACES J*, vol. 24, pp. 541–549, 2009.

[2] L. C. V. Atta, "Electromagnetic reflector," U.S. Patent 2 908 002, October 6, 1959.

[3] P. W. Howells, "Intermediate frequency side-lobe canceller," U.S. Patent 3 202 990, August 24, 1965.

[4] P. Howells, "Explorations in fixed and adaptive resolution at GE and SURC," *IEEE Trans. Antennas Propag.*, vol. 24, pp. 575–584, 1976.

[5] B. Widrow *et al.*, "Adaptive antenna systems," *Proc. IEEE*, vol. 55, pp. 2143–2159, 1967.

[6] R. A. Monzingo *et al.*, *Introduction to Adaptive Antennas*, 2nd ed., Raleigh, NC: Scitech Publishing, 2010.

[7] R. T. Compton Jr., *Adaptive Antennas: Concepts and Performance*, Englewood Cliffs, NJ: Prentice-Hall, 1988.

[8] I. S. Reed *et al.*, "Sample matrix inversion technique," Proceedings of the 1974 Adaptive Antenna Systems Workshop, NRL Report 7803, Naval Research Laboratory, Washington, DC, pp. 219–222, Mar. 11–13, 1974, vol. I.

[9] C. A. Baird Jr. and J. T. Rickard, "Recursive estimation in array processing," Fifth Asilomar Conference on Circuits and Systems, Pacific Grove, CA, Nov. 1971, pp. 509–513.

[10] F. B. Gross, *Smart Antennas for Wireless Communications: with MATLAB*, New York: McGraw-Hill, 2005.

[11] R. L. Haupt, "Adaptive nulling with weight constraints," *Prog. Electromagn. Res. B*, vol. 26, pp. 23–38, 2010.

[12] R. L. Haupt *et al.*, "Adaptive nulling using photoconductive attenuators," *IEEE Trans. Antennas Propag.*, vol. 59, pp. 869–876, 2011.

[13] R. L. Haupt and H. L. Southall, "Experimental adaptive nulling with a genetic algorithm," *Microw. J*, vol. 42, pp. 78–89, 1999.

[14] R. Shore, "Nulling a symmetric pattern location with phase-only weight control," *IEEE Trans. Antennas Propag.*, vol. 32, pp. 530–533, 1984.

[15] C. Baird and G. Rassweiler, "Adaptive sidelobe nulling using digitally controlled phase-shifters," *IEEE Trans. Antennas Propag.*, vol. 24, pp. 638–649, 1976.

[16] R. L. Haupt, "Adaptive nulling in monopulse antennas," *IEEE Trans. Antennas Propag.*, vol. 36, pp. 202–208, 1988.

[17] D. R. Morgan, "Partially adaptive array techniques," *IEEE Trans. Antennas Propag.*, vol. 26, pp. 823–833, 1978.

[18] D. J. Chapman, "Partial adaptivity for the large array," *IEEE Trans. Antennas Propag.*, vol. 24, pp. 685–696, 1976.

[19] R. L. Haupt and D. Werner, *Genetic Algorithms in Electromagnetics*, New York: John Wiley & Sons, 2007.

[20] R. L. Haupt, "Phase-only adaptive nulling with a genetic algorithm," *IEEE Trans. Antennas Propag.*, vol. 45, pp. 1009–1015, 1997.

[21] H. Steyskal *et al.*, "Methods for null control and their effects on the radiation pattern," *IEEE Trans. Antennas Propag.*, vol. 34, pp. 404–409, 1986.

[22] R. L. Haupt, and M. Lanagan, "Reconfigurable antennas," *IEEE AP-S Mag.*, vol. 55, pp. 49–61, 2013.

[23] B. Tomasic *et al.*, "The geodesic dome phased array antenna for satellite control and communication—subarray design, development and demonstration," IEEE International Symposium on Phased Array Systems and Technology, Boston, MA, USA, 2003, pp. 411–416.

[24] R. L Haupt, *Antenna Arrays: A Computational Approach*, Hoboken, NJ: John Wiley & Sons, Inc., 2010.

[25] A. Lyons *et al.*, "Antenna pattern measurements of an S-band satellite communications phased array antenna panel," Antenna Measurement Techniques Association Symposium, Denver, CO, October 2011.

[26] J. C. Webber, "The ALMA telescope," Microwave Symposium Digest (IMS), IEEE MTT-S International, Seattle, WA, USA, June 2013.

[27] *About ALMA technology* [Online]. Available: http://www.almaobservatory.org/. Accessed May 27, 2009.

[28] National Radio Astronomy Observatory, *An overview of the very large array* [Online]. Available: http://www.vla.nrao.edu/genpub/overview/. Accessed June 17, 2009.

[29] H. Steyskal *et al.*, "Pattern synthesis for TechSat21—a distributed space-based radar system," *IEEE Antennas Propagat. Mag.*, vol. 45, pp. 19–25, 2003.

[30] J. S. Herd *et al.*, "Reconfigurable microstrip antenna array geometry which utilizes micro-electro-mechanical system (MEMS) switches," U.S. Patent 6 198 438. March 6, 2001.

[31] F. Yang and Y. Rahmat-Samii, "A reconfigurable patch antenna using switchable slots for circular polarization diversity," *IEEE Microw Wireless Compon. Lett.*, vol. 12, pp. 96–98, 2002.

[32] E. R. Brown, "RF-MEMS switches for reconfigurable integrated circuits," *IEEE Trans. Microw Theory Tech.*, vol. 46, pp. 1868–1880, 1998.

[33] S. Nikolaou *et al.*, "UWB elliptical monopoles with a reconfigurable band notch using MEMS switches actuated without bias lines," *IEEE Trans. Antennas Propag.*, vol. 57, pp. 2242–2251, 2009.

[34] G. Wang *et al.*, "A high performance tunable RF MEMS switch using barium strontium titanate (BST) dielectrics for reconfigurable antennas and phased arrays," *IEEE Antennas Wirel. Propag. Lett.*, vol. 4, pp. 217–220, 2005.

[35] P. Bhartia and BAHL, "Frequency agile microstrip antennas," *Microw J*, vol. 25, pp. 67–70, 1982.

[36] N. Behdad and K. Sarabandi, "A varactor-tuned dual-band slot antenna," *IEEE Trans. Antennas Propag.*, vol. 54, pp. 401–408, 2006.

[37] A. R. Weily *et al.*, "A reconfigurable high-gain partially reflecting surface antenna," *IEEE Trans. Antennas Propag.*, vol. 56, pp. 3382–3390, 2008.

[38] H. Mirzaei and G. Eleftheriades, "A compact frequency-reconfigurable metamaterial-inspired antenna," *IEEE Antennas Wirel. Propag. Lett.*, vol. 10, pp. 1154–1157, 2011.

[39] L. Zhang *et al.*, "Analysis and design of a reconfigurable dual-strip scanning antenna," IEEE AP-S Symposium, Charleston, SC, USA, June 2009.

[40] W. Kummer *et al.*, "Ultra-low sidelobes from time-modulated arrays," *IEEE Trans. Antennas Propag.*, vol. 11, pp. 633–639, 1963.

[41] S. Yang *et al.*, "Evaluation of directivity and gain for time-modulated linear antenna arrays," *Microw. Opt. Technol. Lett.*, vol. 42, pp. 167–171, 2004.

[42] L. Manica *et al.*, "Almost time-independent performance in time-modulated linear arrays," *IEEE Antennas Wirel. Propag. Lett.*, vol. 8, pp. 843–846, 2009.

[43] R. Haupt, "Time modulated receive arrays," IEEE AP-S Conference, Spokane, WA, USA, 2011.

[44] P. Rocca and R. Haupt, "Dynamic array thinning for adaptive interference cancellation," EuCAP, Waltham, MA, USA, April 2010.

[45] P. Rocca and R. Haupt, "Dynamic thinning strategy for adaptive nulling in planar antenna arrays," IEEE Phased Array Symposium, Wayland, MA, October 2010, pp. 995–997.

[46] P. Rocca *et al.*, "Interference suppression in uniform linear arrays through a dynamic thinning strategy," *IEEE Trans. Antennas Propag.*, vol. 59, pp. 4525–4533, 2011.

[47] R. L. Haupt, "Adaptive low sidelobe thinning," ICEAA Conference, Turin, Italy, September 2013.

[48] L. Poli *et al.*, "Reconfigurable thinning or the adaptive control of linear arrays," *IEEE Trans. Antennas Propag.*, vol. 61, pp. 5068–5077, 2013.

[49] R. L. Haupt, "Adaptive nulling by variable array thinning," ICEAA Conference, Aruba, August 2014.

[50] R. Haupt, "Reconfigurable thinned arrays," IEEE Radar Conference, Cincinnati, OH, USA, May 2014.

[51] M. R. Frater and M. J. Ryan, *Electronic Warfare for the Digitized Battlefield*, Boston, MA: Artech House, 2001.

[52] S. Chandran, *Advances in Direction-of-Arrival Estimation*, Boston, MA: Artech House, 2006.

[53] R. L. Haupt *et al.*, "Biological beamforming," in *Frontiers of Mathematical Methods in Electromagnetics*, D. H. Werner and R. Mittra, ed. Piscataway, NJ: IEEE Press, 1999, pp. 329–370.

[54] F. Adcock, "Improvement in Means for determining the direction of a distant source of electromagnetic radiation," British patent 1 304 901 919, 1917.

[55] J. Capon, "High-resolution frequency-wavenumber spectrum analysis," *Proc. IEEE*, vol. 57, pp. 1408–1418, 1969.

[56] R. Schmidt, "Multiple emitter location and signal parameter estimation," *IEEE Trans. Antennas Propag.*, vol. 34, pp. 276–280, 1986.

[57] J. P. Burg, "The relationship between maximum entropy spectra and maximum likelihood spectra," *Geophysics*, vol. 37, pp. 375–376, 1972.

[58] C. Pon, "Retrodirective array using the heterodyne technique," *IEEE Trans. Antennas Propag.*, vol. 12, pp. 176–180, 1964.

[59] R. Y. Miyamoto and T. Itoh, "Retrodirective arrays for wireless communications," *IEEE Microw.*, vol. 3, pp. 71–79, 2002.

[60] B. Nair and V. F. Fusco, "Two-dimensional planar passive retrodirective array," *Electron. Lett.*, vol. 39, pp. 768–769, 2003.

[61] M. K. Watanabe *et al.*, "A 2-D phase-detecting/heterodyne-scanning retrodirective array," *IEEE Trans. Microw Theory Tech.*, vol. 55, pp. 2856–2864, 2007.

[62] S. M. Alamouti, "A simple transmit diversity technique for wireless communications," *IEEE J. Sel. Areas Comm.*, vol. 16, pp. 1451–1458, 1998.

[63] M. A. Jensen and J. W. Wallace, "A review of antennas and propagation for MIMO wireless communications," *IEEE Trans. Antennas Propag.*, vol. 52, pp. 2810–2824, 2004.

[64] A. Molisch, *Wireless Communications*, 2nd ed., West Sussex: John Wiley & Sons, 2011.

[65] T. Brown *et al.*, *Practical Guide to the MIMO Radio Channel*, West Sussex: John Wiley & Sons, 2012.

[66] M. S. Davis and A. D. Lanterman, "Coherent MIMO radar: the phased array and orthogonal waveforms," *IEEE Aerosp. Electron. Syst. Mag, Tutorial VII*, vol. 29, pp. 76–91, 2014.

LIST OF SYMBOLS AND ABBREVIATIONS

a	
$\mathbf{A}$	array steering vector
ACPR	adjacent channel power ratio
αdB	attenuation (dB)
ADC	analog to digital converter
AESA	active electronically scanned array
AF	array factor
AF_e	array factor of one subarray of elements
AF_s	array factor due to point sources at subarray ports at one level
ALMA	Atacama Large Millimeter/submillimeter Array
A_p	aperture area
AR	axial ratio
ATF	Advanced Tactical Fighter
A_{uc}	unit cell area
A_{UC}	area of unit cell
AWGN	additive white Gaussian noise
BAW	Bulk acoustic wave
BER	bit error rate
b_n	bin n
BPSK	binary phase shift keying
BST	$Ba_{0.5}Sr_{0.5}TiO_3$

Timed Arrays: Wideband and Time Varying Antenna Arrays, First Edition. Randy L. Haupt.
© 2015 John Wiley & Sons, Inc. Published 2015 by John Wiley & Sons, Inc.

BW	bandwidth
BWinstantaneous	instantaneous bandwidth
c	speed of light
C	correlation matrix
C_0	transmission line capacitance
C_d	tunable distributed capacitance
C_{max}	maximum tuning capacitance
C_{min}	minimum tuning capacitance
C_{noise}	noise correlation matrix
CNR	carrier-to-noise ratio
C_{off}	switch off capacitance
C_s	signal correlation matrix
CSA	current sheet antenna
$C_{s\text{-}noise}$	signal–noise cross-correlation matrix
δ	phase
$\delta(x)$	delta function
δ^a	Random amplitude error
DAC	digital to analog convertor
D_{array}	array directivity
dBc	decibels above carrier
DBF	digital beamforming
$D_{element}$	element directivity
D_{helix}	diameter of loops in helical antenna
δ_{LSB}	LSB phase
D_{max}	maximum aperture diameter
DMI	direct matrix inversion
δ^p	random phase error
δ_s	beam steering phase
D_{thin}	directivity of a thinned array
d_x	element spacing in x-direction
d_y	element spacing in y-direction
E	electric field
ε	error
E_b	energy in one bit
E_b/N_0	energy per bit to noise power spectral density ratio
ECM	electronic counter measures
ε_{eff}	effective dielectric constant
ε_{max}	maximum relative permittivity
ε_{min}	minimum relative permittivity
ε_r	relative permittivity
f	frequency
ϕ	azimuth angle
F	noise factor
f_{Bragg}	Bragg frequency
f_c	carrier frequency

FET	field-effect transistor
ϕ_g	azimuth direction of grating lobe
f_{hi}	highest frequency in the bandwidth
f_{IMP}	IMP frequencies
f_{lo}	lowest frequency in the bandwidth
FOM	figure of merit
FOM_e	figure of merit for tunable dielectric devices
φ_s	azimuth scan
FSS	frequency selective surface
G	gain
g	trace thickness
γ	step size
GA	genetic algorithm
GaAs	gallium arsenide
GaN	gallium nitride
GBT	Greenbank Telescope
γmn	signal weighting
G_r	gain of receive antenna
G_t	gain of transmit antenna
h	height
h	thickness of substrate
H	channel matrix
HBT	heterojunction bipolar transistor
h_{mn}	element of channel matrix
η_t	taper efficiency
I	current
I	identity matrix
IFF	identification friend-or-foe
IMP	intermodulation products
IMP3	third-order intermodulation product
InP	indium phosphide
IQ	in-phase quadrature
k	wavenumber
K	elliptical integral of the first kind
λ	wavelength
L	length
l	length
L_0	transmission line inductance
LDMOS	lateral double-diffused metal oxide semiconductor
LHCP	left-hand circular polarization
λ_{max}	maximum wavelength in the bandwidth
λ_{min}	minimum wavelength in the bandwidth
LMS	least mean square
LNA	low-noise amplifier
LOS	line of sight

LSB	least significant bit
LTCC	low-temperature cofired ceramic
LWA1	Long Wavelength Array Station 1
Mbps	mega bits per second
MEMS	microelectromechanical systems
MEMS	maximum entropy method
MESFET	metal semiconductor field-effect transistors
MIMO	multiple input multiple output
MMIC	monolithic microwave-integrated circuit
MSB	most significant bit
MUSIC	*MU*ltiple *SI*gnal *C*lassification
N	number of elements
N	noise vector
N_0	noise
N_{arm}	number of arms in a spiral
N_{bits}	number of bits
N_e	number of elements in a subarray
NF	noise figure
N_{helix}	number of loops in helical antenna
n_r	relative tunability
N_s	number of subarrays
N_x	number of elements in x-direction
N_y	number of elements in y-direction
OOK	on–off keying
$P_{1\,dB}$	1 dB compression point
P_{3OI}	third order intercept
PAE	power added efficiency
PCB	printed circuit board
P^e	random element failure
PHEMT	pseudomorphic high electron-mobility transistors
P_I	interference power
p_{max}	maximum eigenvalue
P_n	noise power
p_n	nth eigen value
P_{PASS}	power spectral density in pass band
P_r	power received
P_s	signal power
PSD	power spectral density
P_t	power transmitted
PUMA	planar ultrawideband modular antenna
θ	elevation angle
Q	inverse of the device loss
q	signal correlation vector
$\theta_{3\,dB}$	3 dB beamwidth
QFN	quad-flat no-leads

θ_g	
$\theta_{maxscanx}$	maximum scan angle in x-direction
$\theta_{maxscany}$	maximum scan angle in y-direction
QPSK	quadrature phase shift keying
θ_s	elevation scan
R	distance
RFID	radio frequency identification
RHCP	right hand circular polarization
r_{in}	distance from center of spiral to beginning of spiral arm
R_{on}	switch on resistance
r_{out}	distance from center of spiral to end of spiral arm
σ	standard deviation
s	signal
s	gap widths in coplanar waveguide
s	transmitted signal vector
SER	symbol error rate
s_{helix}	spacing between loops in helical antenna
Si	silicon
SiC	silicon carbide
SISO	single input single output
sll	sidelobe level (V)
slldb	sidelobe level (dB)
sll_{max}	maximum sidelobe level
sll_p	peak sidelobe level
SMI	sample matrix inversion
Smn	s-parameter
σ_{noise}	noise standard deviation
SNR	signal to noise ratio
t	time
T	time period
τ	time delay
T/R	transmit/receive
T_d	time delay
t_{fall}	pulse fall time
τ_g	group delay
τ_g	average group delay
τ_{LSB}	LSB of time delay
τ_{max}	maximum time delay
τ_{MSB}	MSB of time delay
t_{off}	pulse off time
t_{on}	pulse on time
t_{rise}	pulse rise time
T_s	switch period
TSA	tapered slot antenna

u	angle projection
u_s	scan angle projection
UWB	ultra wide band
V	voltage or volts
v	tapered slot antenna
$V_{incident}$	incident voltage
VLA	very large array
V_{max}	maximum voltage
V_{min}	minimum voltage
v_n	velocity
$V_{reflected}$	reflected voltage
v_s	angle projection
v_{sat}	saturation voltage
V_T	threshold voltage
ω	radial frequency
W	width
W	microstrip trace width
$\mathbf{w}$	element weight vector
$\mathbf{W}_\lambda$	matrix with eigenvector columns
w_n	complex element weight
w_{opt}	optimum weight vector
WSA	wavelength-scaled array
$\mathbf{X}$	received signal vector
ψ	phase
ζ	Rapp model smoothness factor
Z_A	antenna impedance
Z_c	characteristic impedance
Z_{in}	input impedance
Z_{mn}	mutual impedance between elements m and n

INDEX

Timed Arrays: Wideband and Time Varying Antenna Arrays, First Edition. Randy L. Haupt.
© 2015 John Wiley & Sons, Inc. Published 2015 by John Wiley & Sons, Inc.

CPSIA information can be obtained
at www.ICGtesting.com
Printed in the USA
BVHW011302300621
610889BV00002B/8